Physical Properties of Materials for Engineers

Volume II

Author

Daniel D. Pollock
Professor of Engineering
State University of New York at Buffalo
Buffalo, New York

CRC Press, Inc.
Boca Raton, Florida

Library of Congress Cataloging in Publication Data

Pollock, Daniel D.
Physical properties of materials for engineers.

Includes bibliographies and indexes.
1. Solids. 2. Materials. I. Title.
QC176.P64 620.1′12 81-839
ISBN 0-8493-6200-8 (set) AACR2
ISBN 0-8493-6201-6 (v. 1)
ISBN 0-8493-6202-4 (v. 2)
ISBN 0-8493-6203-2 (v. 3)

Direct all inquiries to CRC Press, Inc., 2000 N.W. 24th Street, Boca Raton, Florida 33431.

International Standard Book Number 0-8493-6200-8 (Complete Set)
International Standard Book Number 0-8493-6201-6 (Volume I)
International Standard Book Number 0-8493-6202-4 (Volume II)
International Standard Book Number 0-8493-6203-2 (Volume III)

Library of Congress Card Number 81-839
Printed in the United States

PREFACE

Many new materials and devices which were designed to possess specific properties for special purposes have become available in the recent past. These have had their origins in basic scientific concepts. Engineers must understand the bases for these developments so that they can make optimum use of available materials and further advance the existing technology as new materials appear. The main objective of this text is to provide engineers and engineering students a unified, elementary treatment of the basic physical relationships governing those properties of materials of greatest interest and utility.

Many texts on solid state physics, written primarily for advanced undergraduate physics courses, make use of sophisticated mathematical derivations in which only the most significant parts are given; the intermediate steps are left to the reader to provide. This makes it difficult for the average engineer to follow and has the effect of discouraging or "turning off" many readers. Other texts are not much more than surveys of "materials science" and provide little insight into the nature of the phenomena.

This text represents an attempt to provide a middle ground between these extremes. It is designed to explain the origin and nature of the most widely used physical properties of materials to engineers; thus, it prepares them to understand and to utilize materials more effectively. It also may be used as a textbook for senior undergraduate and first-year graduate students.

Practicing engineers will find this text helpful in getting up to date. Readers with some familiarity with this field will be able to follow the presentations with ease. Engineering students and those taking physics courses will find this book to be a useful source of examples of applications of the theory to commercially available materials as well as for uncomplicated explanations of physical properties. In many cases alternate explanations have been provided for clarity.

An effort has been made to keep the mathematics as unsophisticated as possible without "watering down" or distorting the concepts. In practically all cases only a mastery of elementary calculus is required to follow the derivations. All of the "algebra" is shown and no steps in the derivations are considered to be obvious to the reader. Explanations are provided in cases where more advanced mathematics is employed The problems have been designed to promote understanding rather than mathematical agility or computational skill.

The introductory chapters are intended to span the gap between the classical mechanics, which is familiar to engineers and engineering students, and the quantum mechanics, which usually is unfamiliar. The limitations of the classical approach are shown in elementary ways and the need for the quantum mechanics is demonstrated. The quantum mechanics is developed directly from this by the use of uncomplicated examples of various phenomena. The degree to which the quantum mechanics is presented is sufficient for the understanding of the physical properties discussed in the subsequent chapters; it also provides a sound basis for more advanced study.

Introductory sections are given which guide the reader to the topic under consideration. The basic physical relationships are provided. These are drawn from concepts and properties which are known to those with engineering backgrounds; they lead the reader into the topic of interest. In some cases small amounts of material are repeated for the sake of clarity and convenience. Some topics, frequently presented as separate chapters in physics texts, have been incorporated in various sections in which they are directly applicable to materials. Lattice dynamics is one of the subjects treated in this way. Where appropriate, sections covering the properties of commercially available materials are included and discussed. This approach provides more comprehensive presentations which can be readily followed and applied by the reader.

Since this text is intended for readers with engineering backgrounds, some of the topics often presented in solid state physics books have been omitted. The reader is, however, provided with a suitable foundation upon which to pursue such topics elsewhere. The fundamentals of solid state physics are indispensable to the understanding of the properties of materials; these have been retained. Thus, the approach and content of this text are unique in that they include the properties and applications as well as the theory of those major types of real materials which are most frequently employed by engineers. This is rarely, if ever, done in current physics texts.

On the other hand, important subjects marginally included, or omitted, from many physics texts have been incorporated. Chapter 6 (Electrical Resistivities and Temperature Coefficients of Metals and Alloys) and Chapter 7 (Thermoelectric Properties of Metals and Alloys) are good examples of this. These chapters are unique in that similar material does not appear in any text of which I am aware. Sections of these chapters include the basic physical theories and their relationships to phase equilibria as well as their application to the design of alloys with special sets of electrical properties and to the explanations of the properties of commercially available alloys. The very wide use of these types of alloys makes it necessary that engineers thoroughly understand the mechanisms responsible for their optimum applications and their limitations. Other topics of primary importance to engineers which are normally included in solid state physics texts also have been incorporated.

The background required for this text includes elementary calculus, first-year, college-level physics and chemistry, and one course in physical metallurgy or materials science. Information required beyond these levels has been incorporated where needed. This makes it possible to accommodate the needs of readers where there is a wide range of background and capability; it also permits self-study.

The first five chapters introduce, explain, and develop the modern theory of solids; these are considered to constitute the minimum basis for any text of this type. Various other sections, or chapters, may then be studied, depending upon the interests of the reader and the emphasis desired. One combination of topics could be selected by electrical engineers, another set by metallurgical engineers, still another group by mechanical engineers, etc. Courses in materials engineering could be organized in similar ways. It should be noted, however, that all of the major topics included in this text represent physical properties employed by, and of significance to, most engineers at some time during their careers.

Note should be made that the units used in each of the topics are those currently employed by engineers working with materials in that area. The use of a single system of units would be counterproductive. Means for conversions to other units are given in the text for convenience and in the appendix.

I wish to express my deep appreciation to two of my former teachers for the insights and approaches to solid state phenomena which they provided early in my career. Professor C. W. Curtis, of Lehigh University, and Dr. F. E. Jaumot, then associated with the University of Pennsylvania, have been continuing sources of inspiration. In addition, some of the illustrations given in Dr. Curtis' lectures have served, with his permission, as models for the equivalents given here. Similarly, I am indebted to Dr. Jaumot for permission to use his clear approaches to reciprocal space, Brillouin zone theory, and the elementary theory of alloy phases as a basis for those used here.

I am deeply grateful to the American Society for Testing and Materials for permission to condense the contents and to use the illustrations from the monograph, *The Theory and Properties of Thermocouple Elements,* STP492, 1971, written by the author. This material is presented in Chapter 7 (Volume II).

Acknowledgment is also made of the assistance provided by Mr. James Stewart for his cooperation and assistance in the preparation of the illustrations.

I am very grateful to Donna George for her unfailing patience and help in typing the manuscript.

Credits are given with the individual tables and figures.

Daniel D. Pollock

PHYSICAL PROPERTIES OF MATERIALS FOR ENGINEERS

Daniel D. Pollock

Volume I

Beginnings of Quantum Mechanics
Waves and Particles
The Schrödinger Wave Equation
Thermal Properties of Nonconductors
Classification of Solids

Volume II

Electrical Resistivities and Temperature Coefficients of Metals and Alloys
Thermoelectric Properties of Metals and Alloys
Diamagnetic and Paramagnetic Effects
Ferromagnetism

Volume III

Physical Factors in Phase Formation
Semiconductors
Dielectric Properties
Useful Physical Constants
Conversion of Units

TABLE OF CONTENTS

Volume II

Chapter 6

ELECTRICAL RESISTIVITIES AND TEMPERATURE COEFFICIENTS OF METALS AND ALLOYS

Normal metals and some of their alloys are the best conductors because their nearly free electrons only partly fill the Brillouin zones. These electrons have very high velocities and, for elemental metals, have relatively long mean free paths in the lattice (see Section 5.6.3, Chapter 5, Volume I). At the same time, the ions in the lattice oscillate about their equilibrium positions, as discussed in Chapter 4, Volume I. The oscillating ions obstruct the motion of the electrons and limit their mean free paths and relaxation times. These electron-ion "collisions" are known as scattering. As the temperature increases, the number of phonons increases and more effective scattering of the electrons occurs; this offers resistance to the flow of electrons. In terms of Equation 5-64, $L(E_F)$ decreases and $V(E_F)$ only changes slightly, so the relaxation time, $\tau(E_F)$, decreases with increasing temperature. This diminishes the electrical conductivity and increases its reciprocal, the electrical resistivity, as a function of temperature.

Both the electrons and the ions in the lattice, therefore, affect the electrical resistivity and its behavior as a function of temperature (temperature coefficients of resistivity). As would be expected, these properties change with alloying and reflect fundamental changes in the solid state. In fact, the early investigators of phase equilibria (Tamman and Kurnakov, to cite two) made extensive use of such properties in their investigations. These properties have continued to be employed to monitor and/or detect changes in the solid state.

The engineering applications of alloys for electrical components are extensive. These include such applications as thin metallic films in integrated circuits, precision, wire-wound resistors and potentiometers, resistance thermometers, and heating elements for ovens and furnaces, to cite a few.

The ideas presented here will provide a basis for understanding the solid-state phenomena reflected by these properties and explanations of the behaviors of commercially available alloys.

6.1. ELECTRICAL RESISTIVITY (CONDUCTIVITY), DILUTE ALLOYS

In the case of normal elemental metals, the Brillouin zone is partialy filled and, using the spherical approximation for the Fermi level, may be considered as not affecting the electron transport process. Thus, it does not affect the electrical conductivity of such metallic solids. The electrical resistivity must, therefore, result from other interactions.

Almost all of the elemental metals show extended ranges of virtually linear resistivity-vs.-temperature behavior, at temperatures above about 0.2 of the Debye temperature. The normal elements show temperature coefficients of resistivity of about 0.4%/°C. Values of this property for the transition elements may be higher than this and can range up to about 0.7%/°C. for very pure nickel (see Table 6-1). This relative uniformity of behavior permits the generalization of the electrical resistivity of most metals by means of a "universal" curve such as is shown in Figure 6-1. The behavior of superconductors is discussed in Section 6.8. Here the data, ϱ_T, are normalized with respect to the electrical resistivity at θ_D. At cryogenic temperatures, very much lower than θ_D, ϱ varies as T^5. Electron-electron scattering is important in this range. The T^5 behavior may be altered by impurities. Some metals may show minima at these very low temperatures.

Table 6-1
TEMPERATURE COEFFICIENTS OF ELECTRICAL RESISTIVITY OF SOME SELECTED METALS AND ALLOYS

Typical composition (wt%)	Δϱ/ϱΔT (Ω/Ω/°C×10⁶)	Typical composition (wt%)	Δϱ/ϱΔT (Ω/Ω/°C×10⁶)
Al (99.99)	+ 4290 (at 20°C)	78 Cu-22 Ni	+ 300 (0—100°C)
Cu (99.99)	+ 4270 (0—50°C)	55 Cu-45 Ni	± 40 (20—100°C)
Au (99.999)	+ 4000 (0—100°C)		
Fe (99.94)	+ 5000 (at 20°C)	87 Cu-13 Mn	± 15 (15—35°C)
Mo (99.9)	+ 3300 (at 20°C)	83 Cu-13 Mn-4 Ni	± 10 (15—35°C)
Pt (99.99)	+ 3920 (0—100°C)		
Ag (99.99)	+ 4100 (at 20°C)	99.8 Ni	+ 6000 (20—35°C)
Ta (99.96)	+ 3820 (0—100°C)	71 Ni-29 Fe	+ 4500 (20—100°C)
		80 Ni-20 Cr	+ 85 (−55 to 100°C)
W (99.9)	+ 4500 (at 20°C)	75 Ni-20 Cr-3 Al + Cu or Fe	± 20 (−55 to 100°C)
98 Cu-2 Ni	+ 1500 (0—100°C)		
94 Cu-6 Ni	+ 800 (0—100°C)	60 Ni-16 Cr-24 Fe	+ 150 (20—100°C)
89 Cu-11 Ni	+ 400 (0—100°C)	35 Ni-20 Cr-45 Fe	+ 350 (20—100°C)

Note: Also refer to Table 5-6, Chapter 5, Volume I.

From Lyman, T., Ed., *Metals Handbook, Properties and Selection,* Vol. 1, 8th ed., American Society for Metals, Metals Park, Ohio, 1961, 798. With permission.

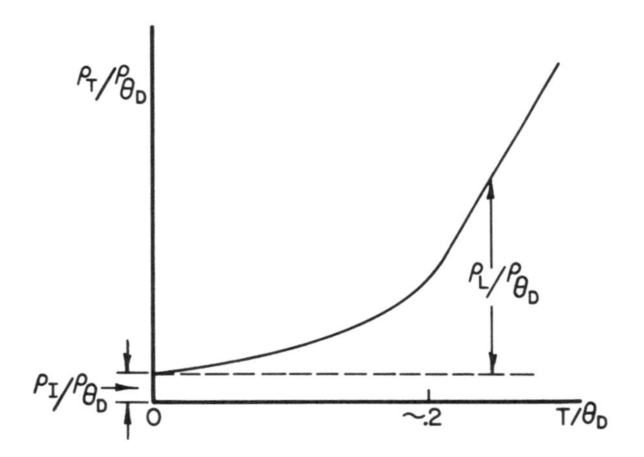

FIGURE 6-1. "Universal" curve for the electrical resistivity of normal metals as a function of temperature relative to the Debye temperature.

The intercept at the ordinate gives clues respecting some of the factors affecting this property. For example, it is found that both the impurity content and, to a much lesser extent, the degree of crystalline imperfection of the materials, significantly affect this parameter in addition to the electron-electron scattering. The magnitude of the intercept diminishes as the purity and crystalline perfection improve. The International Practical Temperature Scale makes use of this in the definition of the platinum reference material in terms of its residual resistance ratio. This is given as an average value of $R_{273}/R_o \simeq 3500$.

The electrical properties of the alloys discussed in subsequent sections will be considered at temperatures well above those where T^5 behavior is present. This will simplify the analyses, since the lattice vibrations involved may be treated more simply.

6.2. MATTHIESSEN'S RULE

Matthiessen (1862) showed that the slopes of the curves of resistivity vs. temperature ($\Delta\varrho/\Delta T$) of well-annealed, very dilute, solid-solution alloys containing about 3 at.% or less, of alloying elements, were the same as that of the annealed, pure (unalloyed), base element. This behavior is shown in Figure 6-2. In other words,

$$\frac{\Delta\rho\,(C_A)}{\Delta T} = \frac{\Delta\rho\,(C_B)}{\Delta T} = \frac{\Delta\rho\,(C_o)}{\Delta T} \tag{6-1}$$

in which C_o denotes the pure metal and C_A and C_B are the compositions of two binary alloys. This behavior is the basis for Matthiessen's Rule. It now is generally expressed as

$$\rho = \rho_I + \rho_L \tag{6-2}$$

for a given temperature, where ϱ is the resistivity of the metal or alloy, ϱ_I is that part of the resistivity induced by the presence of impurities and ϱ_L is the component of the resistivity caused by the scatter of the electron waves by phonons in the lattice.

It was noted previously that the intercept of Figure 6-1 was primarily affected by the impurity content. This is the equivalent of ϱ_I, which is also known as the residual resistivity. The linear portion of the curve, the lattice component (ϱ_L), is also known as the ideal resistivity. The imperfection contribution to the residual resistivity also affects ϱ_L for $T \geqslant \theta_D$, provided that annealing does not occur. Another way of stating Equation 6-2 is to say that the resistivity of a metal or an alloy is the sum of the electron scattering effects of these two factors. This frequently is expressed on the basis that the relaxation time is a function of the probability that an electron will be scattered. The total scattering of an electron, then, can be approximated simply in terms of Equations 6-2 and 5-63 as

$$\rho = \frac{m^*}{n(E_F)e^2\tau(E_F)} = \frac{m^*}{n(E_F)e^2}\left[\frac{1}{\tau(E_F)_I} + \frac{1}{\tau(E_F)_L}\right] \tag{6-3}$$

where the subscripts are the same as those for Equation 6-2. From this, the total relaxation time is found to be

$$\frac{1}{\tau(E)_F} = \frac{1}{\tau(E_F)_I} + \frac{1}{\tau(E_F)_L} \tag{6-4}$$

Here, all of the impurity and lattice effects are lumped in the two parameters. This is somewhat oversimplified but is sufficient for present purposes (see Section 6.3).

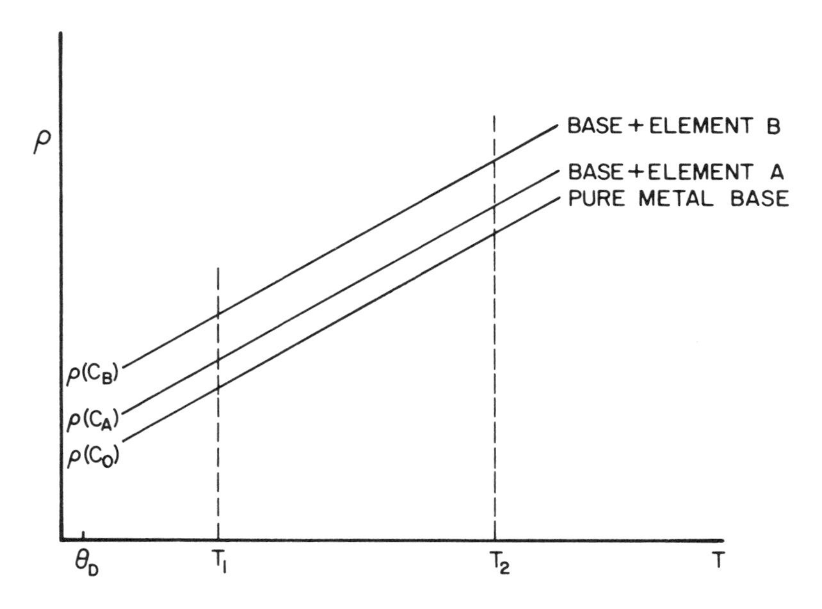

FIGURE 6-2. Electrical resistivity as a function of temperature for annealed dilute, random, binary solid solutions.

It would appear to be inconsistent with Equation 6-4 that several dilute alloys, each containing different alloying elements, should have the same slope as the base, or pure, metal. The differences in their residual resistivities can be understood simply by considering that $\tau(E_F)_I$ varies with the kind and amount of impurity ions and is the predominating factor in Equation 6-3 at very low temperatures. But, why should $\tau(E_F)_L$ be virtually unaffected by the various dilute alloys in a given base metal at elevated temperatures?

In order to answer this question, it is necessary to examine the factors in Equation 5-63 in greater detail in the temperature range close to, or above θ_D. Under this condition, both the Einstein and Debye models for lattice vibrations give similar and reasonable results (see Chapter 4, Volume I). The Einstein model is used here because of its greater simplicity. Each vibrating ion in the lattice may be considered to be a simple harmonic oscillator with a constant frequency as in Section 4.1.2, Chapter 4, Volume I. The potential energy of such a particle is given by

$$\text{P.E.} = \frac{1}{2} M\omega^2 \overline{x}^2 = \frac{1}{2} M \cdot 4\pi^2\nu^2 \overline{x}^2 \tag{6-5}$$

in which ω and ν have the same meanings as in Chapter 4, Volume I, M is the mass of the ion and $\overline{x}$ is its average displacement. The average potential energy is one-half of the thermal energy for $T > \theta_D$. This may be equated with Equation 6-5 to give

$$2\pi^2 M\nu^2 \overline{x}^2 = \frac{1}{2} k_B T \tag{6-6}$$

An electron with a given mean free path, $L(E_F)$, has a 100% probability of being scattered by an oscillating ion. Thus, the scattering cross-section of the ion, $A(E_F)$, may be defined by

$$n(E_F)L(E_F)A(E_F) = 1$$

$$\text{or,} \quad A(E_F) = \frac{1}{n(E_F)L(E_F)} \tag{6-7}$$

where $n(E_F)$ is the number of electrons per unit volume which enter into the conduction process. $A(E_F)$ is an effective area swept out by an oscillating ion and will vary as $\overline{x}^2$; both have the same dimensions. Equation 6-7 is substituted into Equation 6-6 to include this scattering. This is done by starting with the inclusion of $A(E_F)$ in Equation 6-6 to get

$$2\pi^2 M\nu^2 A(E_F) \simeq \frac{1}{2} k_B T$$

When the substitution is made for $A(E_F)$,

$$\frac{1}{n(E_F)L(E_F)} = \frac{k_B T}{4\pi^2 M\nu^2}$$

This is solved to give

$$L(E_F) = \frac{4\pi^2 M\nu^2}{n(E_F)k_B T} \tag{6-8}$$

The numerator and denominator of the fraction are multiplied by $k_B h^2$, and

$$L(E_F) = \frac{4\pi^2 k_B}{n(E_F)h^2} \cdot \frac{Mh^2\nu^2}{k_B^2 T} \tag{6-9}$$

Since Equation 6-5 is applicable at, or near θ_D, Equation 4-101 may be used in Equation 6-9. This simplifies the relationship giving

$$L(E_F) = \frac{4\pi^2 k_B}{n(E_F)h^2} \cdot \frac{M\theta_D^2}{T} \tag{6-10}$$

Equation 6-10 now may be substituted into Equation 5-63a:

$$\sigma = \frac{n(E_F)e^2 L(E_F)}{m^* v(E_F)} = \frac{n(E_F)e^2}{m^* v(E_F)} \cdot \frac{4\pi^2 k_B}{n(E_F)h^2} \cdot \frac{M\theta_D^2}{T}$$

Simplification results in

$$\sigma = \frac{4\pi^2 k_B e^2}{h^2 m^* v(E_F)} \cdot \frac{M\theta_D^2}{T} \tag{6-11}$$

It will be recognized that the denominator of Equation 6-11 contains the product $m^*v(E_F)$; this is the momentum of an electron at the Fermi surface, $p(E_F)$. In normal metals, the Brillouin zones either are partly filled, or zone overlap occurs. In each of these cases, the electrons may be approximated as being free (see Sections 5.8.2 and

5.8.4, Chapter 5, Volume I). The momentum of such an electron may be given by $p(E_F) = (2m^*E_F)^{1/2}$. Now, using Equation 3-25 for the energy of a free electron, slightly modified to fit this case, and assuming a spherical Fermi surface which is unaffected by the zone walls, results in

$$p(E_F) = (2m^*E_F)^{1/2} = \left[2m^* \frac{h^2\bar{k}(E_F)^2}{8\pi^2 m^*}\right]^{1/2} = \left[\frac{h^2\bar{k}(E_F)^2}{4\pi^2}\right]^{1/2}$$

$$= \frac{h\bar{k}(E_F)}{2\pi} \qquad (6\text{-}12)$$

This expression is substituted for the momentum term in Equation 6-11 to give

$$\sigma = \frac{4\pi^2 k_B e^2}{h^2} \cdot \frac{2\pi}{h\bar{k}(E_F)} \cdot \frac{M\theta_D^2}{T} \qquad (6\text{-}13)$$

or, in terms of the electrical resistivity,

$$\rho = \frac{h^3\bar{k}(E_F)T}{8\pi^3 e^2 M\theta_D^2 k_B} = \frac{\hbar^3\bar{k}(E_F)T}{e^2 M\theta_D^2 k_B} \qquad (6\text{-}14)$$

The rate of change of the resistivity with temperature is obtained from the derivative of Equation 6-14; this is

$$\frac{\partial\rho}{\partial T} = \frac{\hbar^3\bar{k}(E_F)}{e^2 M\theta_D^2 k_B} \qquad (6\text{-}15)$$

It will be recalled that E_F is a function of the electron:ion ratio, Equation 5-24. The small alloying additions (<3 at.%) result in negligibly small changes in E_F when compared with that of the pure metal. Consequently, $\bar{k}(E_F)$ remains essentially unchanged. All of the other factors in Equation 6-15 may be treated as constants; M and θ_D may be considered to be unchanged by the small quantities of alloying ions in solid solution in the metal. Therefore, since all of the factors in Equation 6-15 are constant, the slopes of the resistance-temperature curves of these very dilute alloys are constant and must be the same as that of the pure base metal as given in Equation 6-1 and shown in Figure 6-2; this verifies Matthiessen's observations.

The residual resistivities, being primarily a result of the alloy content, shift the curves and account for the differences in the resistivities of the various dilute alloys at a given temperature. These explain the differences between $\varrho(C_O)$,$\varrho(C_A)$, and $\varrho(C_B)$ at a given temperature in Figure 6-2.

It also will be observed that both Equations 6-14 and 6-15 can account for anisotropic variations of these properties in crystals. This behavior is taken into consideration by the inclusion of the wave vector in both equations.

When very accurate measurements of the resistivity are made over a temperature range of several hundred degrees, it is found that very small departures from linearity exist. These small deviations occur above and below the nominal linear relationship and are S-shaped in character. This behavior is taken into consideration in the calibration of highly accurate resistance thermometers. The linear function derived here is very good for the majority of engineering applications.

6.3. ELECTRICAL RESISTIVITY, HIGHER CONCENTRATION BINARY ALLOYS

Two-component, random solid solutions of normal metals, with alloy concentrations greater than 3 at.%, represent the least complicated case to illustrate the mechanisms responsible for the resistivities of the alloys containing larger amounts (>3 at.%) of alloying elements. The assumption again is made that the Brillouin zone is incompletely filled and that the spherical Fermi surface is sufficiently far from the zone boundaries as to be unaffected by them. The validity of this approximation for solid solutions is confirmed in Section 10.6.6.1, Chapter 10, Volume III. However, the other approximations employed to derive Equation 6-14, to explain the properties of very dilute alloys, are not valid in the present case. The mechanisms responsible for the resistivities of the more concentrated solid solutions are considerably more complex than those noted for the dilute solid solutions.

One way to explain the behavior of this class of alloys is to begin by examining Equation 5-63. The most significant factor in this equation, with respect to the alloys being considered here, is the relaxation time. Beginning with Equation 5-64, and using Equation 6-7, this factor can be re-expressed in the following way:

$$\tau(E_F) = \frac{L(E_F)}{V(E_F)}$$

$$L(E_F) = \frac{1}{n(E_F)A(E_F)_T}$$

so

$$\tau(E_F) = \frac{1}{n(E_F)V(E_F)A(E_F)_T} \tag{6-16}$$

Here $A(E_F)_T$ is defined as the total scattering cross-section; this is discussed more completely below. When Equation 6-16 is substituted into Equation 5-63a, it is found that

$$\sigma(E_F) = \frac{n(E_F)e^2L(E_F)}{m^*} = \frac{n(E_F)e^2}{m^*} \cdot \frac{1}{n(E_F)V(E_F)A(E_F)_T}$$

This reduces to

$$\sigma(E_F) = \frac{e^2}{m^*V(E_F)A(E_F)_T} \quad \text{or} \quad \rho = \frac{m^*V(E_F)A(E_F)_T}{e^2} \tag{6-17}$$

As noted earlier, solid solutions with alloy concentrations well within the limits of solid solubility, may be approximated as having spherical Fermi surfaces which are unaffected by proximity to the Brillouin zone boundaries. Thus, m* may be considered to be constant as long as this condition exists.

Alloy additions to the base metal will increase E_F, since according to Equation 5-24 it is a function of the electron:ion ratio. However, the percentage increase in E_F will be relatively small, since E_F is quite large prior to any alloying (see Table 5-2, Chapter 5, Volume I). So it may be assumed that such alloy additions will induce minor changes in $V(E_F)$; such changes may be neglected in this approximation.

The remaining factor in Equation 6-17 most significantly affected by ions in random solid solution is $A(E_F)_T$. The changes in the other factors previously discussed are relatively small compared to those induced in the total scattering cross-section. This factor includes both of the scattering effects of the host and impurity ions, because each of these is contained in $\tau(E_F)$ (Equation 6-4). This is implicit in Equations 5-63 and 5-64, upon which this analysis is based. $A(E_F)_T$ contains another source of scattering which was not included in Equation 6-4; its effects were small and did not require consideration in the case of the very dilute alloys. In the case of the more concentrated alloys, however, changes in the periodic potential within the lattice become very significant; they cause a high degree of scattering and must be taken into account. So, in order to provide a more accurate and complete relationship, Equation 6-4 must be rewritten, as follows, to include this factor:

$$\frac{1}{\tau(E_F)} = \frac{1}{\tau(E_F)_I} + \frac{1}{\tau(E_F)_L} + \frac{1}{\tau(E_F)_e} \qquad (6\text{-}18)$$

The last term gives that component of the total relaxation time due to changes in the periodic potential of the lattice; it results from relatively large numbers of alloying ions, with different charges than that of the host ions, and which occupy sites in the host lattice.

In a way corresponding to the derivation of Equations 6-4 and 6-18, the total scattering cross-section may be approximated to be

$$A(E_F)_T = A(E_F)_L + A(E_F)_e \qquad (6\text{-}19)$$

If the lattice were perfect, and contained none of the usual lattice imperfections, including the distortions resulting from impurity ions, the only resistance to electron flow would be that caused by the thermal oscillations of the host ions; the resistivity would be quite low at relatively low temperatures. However, all of these sources of lattice distortion are present in real alloys; they disrupt the spatial periodicity of the lattice. All such scattering effects are included in $A(E_F)_L$.

Alloying ions in solution in a host lattice usually have different charges from those of the host ions. These differences change the periodic electrical potential of the lattice and contribute to electron scattering. This effect is contained in $A(E_F)_e$.

Rather than attempt to determine the total effects of these two scattering parameters, it is more convenient to consider the *changes* in each of these factors relative to the annealed, unalloyed metal at a given temperature. Such changes are obtained more readily by experiment and explained more simply by analyses. So for these reasons, Equation 6-19 is usually expressed as isothermal changes given by

$$\Delta A(E_F)_T = \Delta A(E_F)_I + \Delta A(E_F)_e \qquad (6\text{-}19a)$$

The electron scattering effects of electrical charge differences between the ions will be approximated first. The host ions each have a charge of $Z_\alpha e$, where Z_α is the valence of the atom and e is the charge on an electron. Each alloy, or foreign, ion has a charge of $Z_\beta e$, where Z_β is its atomic valence. The difference in the charge between an impurity ion and a host ion is $(Z_\beta - Z_\alpha)e$. The electron scattering will be proportional to the square of the difference of these charges, or

$$\Delta A(E_F)_e \propto (Z_\beta - Z_\alpha)^2$$

per impurity atom. On an atomic percent basis,

$$\Delta A(E_F)_e = k_a(Z_\beta - Z_\alpha)^2 \Delta C \quad (6\text{-}20)$$

in which k_a is a constant of proportionality and ΔC is atom percent of the foreign ions present in solid solution in the host lattice.

The effect of size differences may be expressed in terms of Vegard's Rule (see Chapter 10, Volume III). This is accurate for many solid solutions; it was observed first in the behavior of ionic crystals. This is shown schematically in Figure 6-3. Here element D represents the host lattice in which ions of elements E or F are present on random lattice sites. If element E has a larger ionic "diameter" than D, the lattice parameter of the solid solution would be expected to increase. If the ionic "diameter" of element F is smaller than that of D, the lattice parameter of the solution would be expected to decrease with increasing amounts of F. In either event, an alloy ion on a host lattice site will distort, or "warp," a volume of the host lattice which surrounds it. This results from the fact that no two species of ions have exactly the same "diameters." Larger alloy ions cause a bulging, while smaller ions bring about a contraction of the surrounding host lattice. Thus, the kind and amount of distortion which occurs depend upon the relative sizes of the host and alloy ions; this governs the changes in the spatial periodicity of the host lattice. Such effects are lumped with those of any other imperfections (such as dislocations, grain boundaries, vacancies, interstitials, etc.) normally present in the host lattice. All of the factors present within the host lattice which contribute to the scattering of electrons are included in $\Delta A(E_F)_L$.

The change in the lattice parameter (periodicity), using Vegard's Rule, can be expressed by

$$\Delta a_o = \alpha \Delta C \quad (6\text{-}21)$$

where α is a constant of proportionality and a_o is the lattice parameter. The effect of this behavior on the scattering cross-section is given by

$$\Delta A(E_F)_L = k_b \Delta C \quad (6\text{-}22)$$

where the constant, k_b, includes the effects of imperfections in the host lattice. Equations 6-20 and 6-22 can be added, as indicated by Equation 6-19a, to give

$$\Delta A(E_F)_T = k_b \Delta C + k_a(Z_\beta - Z_\alpha)^2 \Delta C$$

or

$$\Delta A(E_F)_T = [k_b + k_a(Z_\beta - Z_\alpha)^2] \Delta C \quad (6\text{-}23)$$

Then, since Equations 6-19 and 6-19a are directly related to Equations 6-3 and 6-4, Equation 6-23 is substituted into the differential of Equation 6-17 to give

$$\Delta\rho = [k_2 + k_1(Z_\beta - Z_\alpha)^2] \Delta C \quad (6\text{-}24)$$

This is the Linde equation (1931) for the change in the electrical resistivity of an annealed, binary alloy as a function of the alloying element present in an annealed, solid solution at a constant temperature.

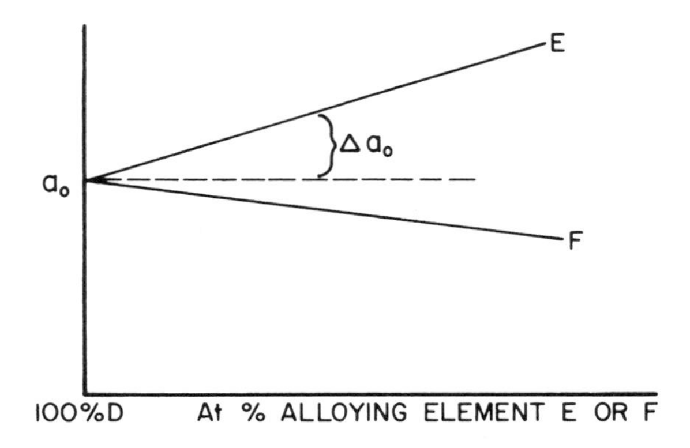

FIGURE 6-3. Change in lattice parameter, a_o, as a function of the species of ions in annealed, random, binary solid solutions at a given temperature.

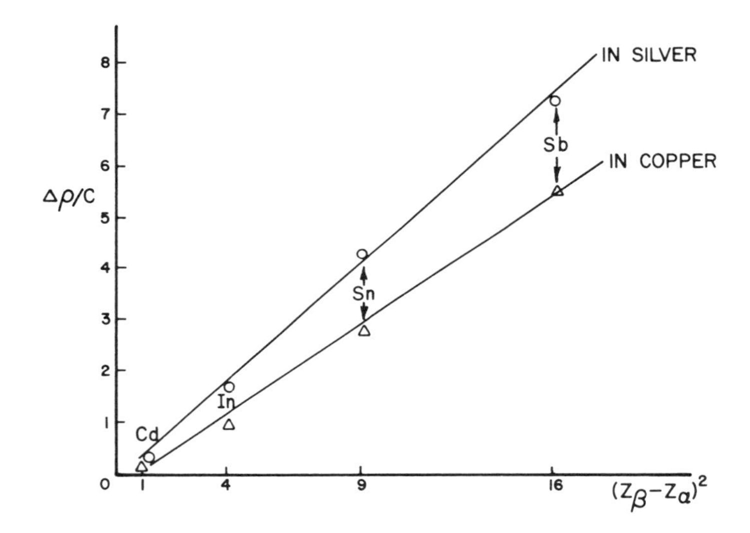

FIGURE 6-4. Effects of ionic charge upon the change in resistivity per atom percent of alloying elements in annealed, random, binary solid solutions at 18°C (see Table 6-2).

If the quantity $\Delta\varrho/\Delta C$ is plotted against $(Z_\beta - Z_\alpha)^2$, the resultant, for solutes of a given period in the Periodic Table, should be linear (Figure 6-4). The slope of the line is k_1 and the intercept is k_2. The constants k_1 and k_2 are valid for solutes from a given period in the Periodic Table, including the transition elements of that period. Where the solutes and solvent ions are of the same period, k_2 becomes quite small. Different values of k_1 and k_2 hold for solute elements of different periods in a given base. The general behavior is shown in Figure 6-4 and Table 6-2.

When more than one alloying element is present in solution in the base, Equation 6-24 holds for each of the constituents, provided that the limit of solid solubility is not exceeded by the combined alloy additions. The effects of each of the several alloying elements are additive and the change in the resistivity of the base metal, at a constant temperature, is given by

$$\Delta\rho = \Sigma\Delta\rho_i \qquad (6\text{-}25)$$

where $\Delta\varrho_i$ is the resistivity change caused by one of the alloying elements. Again, care must be taken to ensure that all of the alloy components are in solution in the base and that the alloy is thoroughly annealed.

The Linde equation (Equation 6-24), used in conjunction with Equation 6-25, constitutes one of the most important bases for the design of commercial engineering alloys for special electrical applications. These equations also give insight into the properties of commercially available alloys (also see Section 6.4).

The understandings provided by Equations 6-24 and 6-25 are somewhat limited because they are restricted to isothermal properties. Additional knowledge may be obtained by applying Equation 6-14 to multicomponent alloys in a way analogous to that used to obtain Equation 6-3. This results in

$$\rho = \frac{\hbar^3 T}{e^2\theta_D^2}\left[\frac{\bar{k}(E_F)_0}{M_0} + \frac{\bar{k}(E_F)_1}{M_1} + \frac{\bar{k}(E_F)_2}{M_2} + \ldots\right] \qquad (6\text{-}26)$$

where the subscript zero refers to the host ions, and the integers each represent one of the several alloying ions. Here, θ_D is the Debye temperature of the given, multicomponent alloy. Equation 6-26 may be simplified to read

$$\rho = \frac{\hbar^3 T}{e^2\theta_D^2}\,\bar{k}(E_F)_T \sum_{i=0}^{n}\frac{1}{M_i} \qquad (6\text{-}27)$$

In this case, $\bar{k}(E_F)_T$ is the wave vector which results from all of the scattering effects of the alloying and host ions and the imperfections in the host lattice. Where the alloy concentration is high, the scattering effects of lattice defects, such as dislocations, imperfections, grain boundaries, etc. are negligible compared to those caused by alloying. Large numbers of alloy ions in solution decrease the mean free path, or wavelength, of the electrons and increase $\bar{k}(E_F)_T$. This increases the resistivity at a given temperature. The larger the amounts of alloy ions in solution, the greater will be the resistivity at the given temperature, provided that solid solubility is maintained.

Increasing the temperature of such an alloy, with a fixed composition, will have a similar but much smaller effect upon $\bar{k}(E_F)_T$. The umklapp processes are very numerous in the range of temperatures in which Equation 6-27 is applicable (see Section 4.4, Chapter 4, Volume I). Thus, the increase in scattering of electrons by phonons will be small compared to the scattering caused by the alloy ions; the slopes of the curves of resistivity vs. temperature of such alloys would be expected to be small. In fact, a general rule is that the greater the alloy content in random solid solution, the smaller will be the slope of this curve.

In cases in which the alloy content is sufficiently large that increases in temperature cause the Fermi surface to begin to overlap into the next Brillouin zone, the wavelengths, or mean free paths, of the electrons increase slightly. This has the effect of slightly decreasing $\bar{k}(E_F)_T$ as the overlap increases with increasing temperature, thus decreasing the resistivity similarly. Alloys of this kind show a nearly parabolic-type resistivity vs. temperature relationship. These have maxima at the temperatures at which the overlapping of the zones becomes appreciable. Some such alloys also show minima at about 300 to 400°C, where $\bar{k}(E_F)_T$ begins to increase again. The temperatures at which the maxima occur and the curve shapes can be controlled by varying the composition of the alloy (see Section 6.7).

Table 6-2
THE CHANGE IN RESISTIVITY PER ATOM PERCENT OF ALLOYING ELEMENT IN AU, AG, AND CU

Alloying element	Solvent metal		
	Au	Ag	Cu
Cu	0.485[a]	0.068[a]	—
Ni	1.00	—	1.25[a]
Co	6.1	—	6.4
Fe	7.66	—	9.3
Mn	2.41	—	2.83
Cr	4.25	—	—
Ti	14.4	—	—
Ag	0.38	—	0.14
Pd	0.407	0.436	0.89
Rh	4.2	—	4.40
Au	—	0.38	0.55
Pt	1.02	1.59	2.51
Ir	—	—	6.1
Zn	0.96	0.62	0.335
Ga	2.2	2.28	1.40
Ge	5.2	5.52	3.75
As	—	8.46	6.8
Cd	0.64	0.382	0.21
In	1.41	1.78	1.10
Sn	3.63	4.32	2.85
Sb	—	7.26	5.45
Hg	0.41	0.79	1.00
Tl	—	2.27	—
Pb	—	4.64	—
Bi	—	7.3	—

[a] In microhm-cm per atom percent at 18°C.

From Linde, J. O., *Ann. der Phys.*, 5 Folge, Band 15, 239, 1932. With permission.

One other solid-state reaction requires mention here. Where short-range ordering takes place, the calculated value for $\Delta\varrho$ will be larger than the experimental value by about 4%. Where long-range ordering occurs, this difference will be very much greater. This decrease in the resistivity comes about because most of the alloy atoms no longer occupy random lattice sites. This increases the degree of lattice regularity, and effectively decreases $\overline{k}(E_F)_T$ in Equations 6-14 or 6-27. This explains the decrease in resistivity observed when ordering occurs (see Section 6.5.2). Alloys showing this behavior are not normally used in electrical instrumentation.

6.4. TEMPERATURE COEFFICIENTS OF RESISTIVITY

The temperature coefficients of resistivities of alloys must be known if resistance materials are to be used properly in engineering applications. This is the case because Joule or ambient heating will change the temperature of the component, and, consequently, its resistance. Therefore, it is very helpful if this property can be approximated closely. It turns out that this can be done quite simply by the means derived here.

As previously noted, Matthiessen's Rule is valid only for relatively dilute alloys. This may be used to obtain an approximation of the temperature dependence of resistance alloys (>3 At.% alloying elements) starting with Equation 6-1

$$\frac{\Delta\rho\,(C_A)}{\Delta T} = \frac{\Delta\rho\,(C_o)}{\Delta T}$$

Multiplying the left-hand side by $\varrho\,(C_A)/\varrho(C_A)$ and the right-hand side by $\varrho(C_o)/\varrho(C_o)$ one obtains

$$\rho\,(C_A)\left[\frac{\Delta\rho\,(C_A)}{\rho\,(C_A)\Delta T}\right] = \rho\,(C_o)\left[\frac{\Delta\rho\,(C_o)}{\rho\,(C_o)\Delta T}\right] \qquad (6\text{-}28)$$

Since the temperature coefficient of electrical resistivity is defined by

$$\alpha \equiv \frac{\Delta\rho}{\rho\Delta T}\left(\frac{\Omega}{\Omega^{o}C}\right) \qquad (6\text{-}29)$$

Equation 6-28 becomes

$$\rho\,(C_A)\alpha_A = \rho\,(C_o)\alpha_o \qquad (6\text{-}30)$$

The properties of the pure base materials are known so that

$$\rho\,(C_A)\alpha_A = \rho\,(C_o)\alpha_o = K_M \qquad (6\text{-}31)$$

where K_M is the Matthiessen constant.

Equations 6-25 and 6-31 can be employed to predict the electrical resistivity and temperature coefficient of nearly all annealed solid-solution alloys. At any given temperature

$$\rho\,(C_A) = \rho\,(C_o) + \Delta\rho\,(C_A) \qquad (6\text{-}32)$$

Since $\Delta\varrho(C_A)$ is given by Equation 6-24, it can be employed in Equation 6-32 to give the resistivity of the annealed solid-solution alloy. This, in turn, can be substituted into Equation 6-31 to compute its temperature coefficient of resistivity. Calculations for α made in this way give good approximations for alloys for which $\partial\varrho/\partial T$ is linear.

The temperature coefficient of resistivity, α, is very sensitive to the small quantities of impurity elements present in solution in a "pure" metal. It constitutes a very accurate, inexpensive means to determine the purity of metals. This is shown schematically in Figure 6-5. The presence of very small amounts of impurities has a large, negative effect upon α, making it significantly smaller. For example, use was made of this sen-

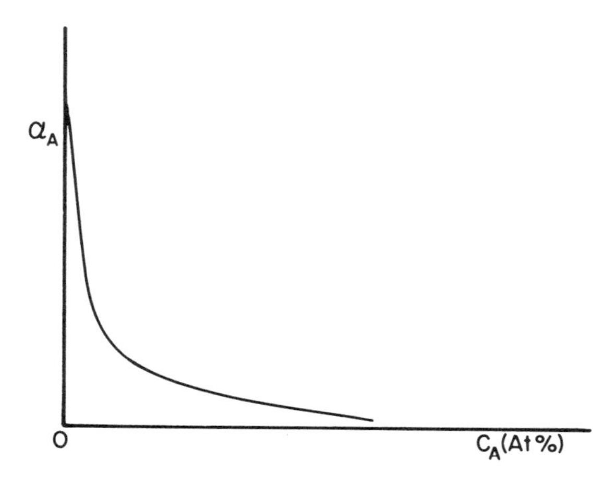

FIGURE 6-5. Temperature coefficient of electrical resistivity (or resistance) as a function of composition of annealed, random, binary solid solutions for a given temperature range.

sitivity to define the composition of the international standard for pure platinum for thermoelectric purposes. Here, the mean value of the platinum (Pt-67) used as a reference material for the International Practical Temperature Scale is defined as possessing a temperature coefficient of electrical resistivity of 0.003269 Ω/Ω/°C from 0 to 100°C. Other pure metallic elements, such as copper and nickel, which are intended for applications in very accurate electrical measuring or sensing devices, such as resistance thermometers, are similarly defined. Such definitions describe degrees of purity far beyond the capacities of most methods of chemical analyses to discriminate. They have the additional advantages of being verified rapidly and inexpensively. This technique, by itself, only can indicate that impurities are present; it cannot provide chemical analyses.

The value of α is lowered by cold working a given alloy. Cold working increases ϱ, and, as may be seen from Equation 6-31, α is diminished accordingly. However, such increases in ϱ are limited (see Section 6.6); this limits the change in α. Data in the literature for the electrical properties of cold worked metals and alloys are sometimes subject to question because stress relief may have occurred at room temperature in the interval before measurement. In addition, the heat generated in the metal during cold working can be sufficient to induce stress relief, recovery, or even recrystallization, if the working is sufficiently severe.

6.5. APPLICATION TO PHASE EQUILIBRIA

Electrical properties were among the earliest means employed to determine the ranges of phases as functions of composition. A series of isothermal plots of these properties, or their changes, vs. composition are still used for such work. The changes in the slopes of such plots indicate phase boundaries as functions of composition and temperature. The use of two of these properties for this purpose is described here.

Equations 6-24 and 6-25 can be employed to verify phaseal relationships both for binary and pseudo-binary systems. Here, isothermal electrical measurements of the equilibrated alloys are made. The only variable is that of composition. Both of the above equations hold only within the limits of solid solubility. Beyond the limits of solid solubility a change in slope occurs because the resistivity then follows the law of

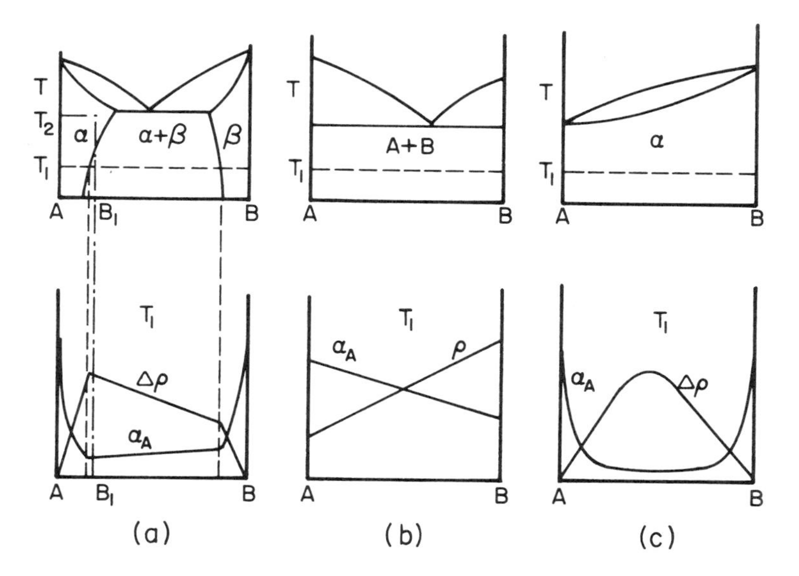

FIGURE 6-6. Relationships between electrical resistivity and temperature coefficient of resistivity as functions of composition for some common types of phase equilibria.

mixtures. This is shown in Figure 6-6a. The hyperbolic behavior noted in Equation 6-29 and Figure 6-5 can be used to verify the resistivity data, but the demarcation between fields is much more difficult. Both methods are applicable in the solid state.

Where complete solid insolubility exists between two components, or phases, the relationships are as shown in Figure 6-6b. Here, a linear relationship exists across the entire diagram because this case represents a mixture.

In the case of isomorphous systems (complete solid solubility) the resistivity is a linear function of composition for about 30% of the range beyond each pure component. It reaches a maximum in the neighborhood of 50% of each component (Figure 6-6c).

On the basis of the foregoing it can be seen that data from well-annealed alloys, equilibrated and measured at various isotherms, can reveal phase fields and boundaries. Of course, this technique is limited to temperatures below the solidus.

These techniques have also been employed in complex alloy systems. Consider an alloy of n components. The alloy base will consist of (n-1) components which are maintained at constant ratios. The composition and properties of this base then constitute one terminal of a pseudo-binary system. Alloy additions of the one component being investigated are made to the complex terminal-alloy base. In employing this approach it is vitally necessary to maintain the (n-1) components of the base very closely to their nominal ratios.

6.5.1. Precipitation Effects

Many engineering alloys which derive desirable sets of mechanical properties from precipitation phenomena must be used with caution, if at all, in electrical applications. The reason for this is that small changes in the precipitates, which may occur over long times, may cause relatively large changes in the electrical properties, even at room temperature.

As previously noted (Section 6.3), the presence of alloying elements in solid solutions increases the electrical resistivity. The precipitation of such ions, whether they are in the form of the pure component, a terminal solid solution, or as an intermetallic phase, would be expected to cause a decrease in the electrical resistivity of the alloy. This frequently does occur, but the resistivity of an equilibrated two-phase alloy may be greater or less than that of one of the terminal solid solution alloys. The slope depends upon the properties of the two alloys of which the two-phase mixture is composed.

Reference to Figure 6-6a shows that precipitation of a second phase from a supersaturated solid solution can either increase or decrease the resistivity of the alloy compared to the terminal solid solutions. These are assumed to be noncoherent precipitates. Here the A-rich, two-phase alloys show a decrease in ϱ. The B-rich alloys show the opposite. Both of these cases assume equilibrium microstructures.

An alloy of composition B_1, in Figure 6-6a, which had been rapidly quenched (see below) from temperature T_2 would represent a metastable nonequilibrium solid solution. Its higher resistivity will diminish with time, the rate depending upon the temperature at which the alloy is maintained. Its properties will vary depending upon the type of precipitate, its precipitation mode, and the degree of departure from equilibrium. For example, small particles precipitating within a grain will impart different effects when compared to similar precipitation at the grain boundaries. In addition, if these particles are smaller than the equilibrium size, the electrical properties of the alloy will change if and when they are allowed to grow so that they approach equilibrium conditions. The composition of these particles also influences the electrical properties of the alloy.

If, however, the precipitates are coherent (have lattices with close registries with the host lattice) in nature, added effects are noted. The mechanisms involved in the precipitation of such phases are considered to follow the Guinier-Preston (G-P) theory (see Figure 6-7). As the incipient particles emerge from solution and grow, they add a large strain component to the lattice which increases $\overline{k}(E_F)_T$. The alloy content of the host lattice becomes depleted slightly. This lattice strain energy overwhelmingly predominates and increases as the G-P zones grow and increase the lattice distortion. This strain component reaches a maximum. At this point the particles start to approach the equilibrium composition and structure and begin to lose coherency with the host lattice. When this occurs the strain energy starts to diminish, the host lattice becomes less distorted, and the resistivity reflects this. As the particles continue to grow they behave increasingly like noncoherent particles until they approach noncoherent behavior and equilibrium composition. They finally become dispersed particles in the host matrix which approach their equilibrium composition.

The detection of maximum coherency is difficult to observe by most methods of detection, including electron microscopy. Resistance, or resistivity, data are ideal for the purpose of determining aging times and temperatures since they readily and economically reflect the precipitation process. Maximum hardness, yield, and tensile strength occur in precipitation-hardening alloys when the number of G-P zones is greatest. This resistometric method is used commercially for the accurate determination of the thermal treatments of such alloys. This is particularly true of aluminum alloys used in aircraft. These alloys frequently are aged for times just beyond that of the maximum shown in Figure 6-7. This is done to improve their resistance to corrosion at the expense of relatively small decreases in strength.

It should be noted that quenching a metallic specimen will also induce lattice strains. The degree of such strains will depend upon the modulus of elasticity of the material as a function of temperature, the magnitude of the temperature differential, and the rate of the quench. Material strained in this way also will show an increase in resistivity; this will be metastable and decrease with time, even at room temperature.

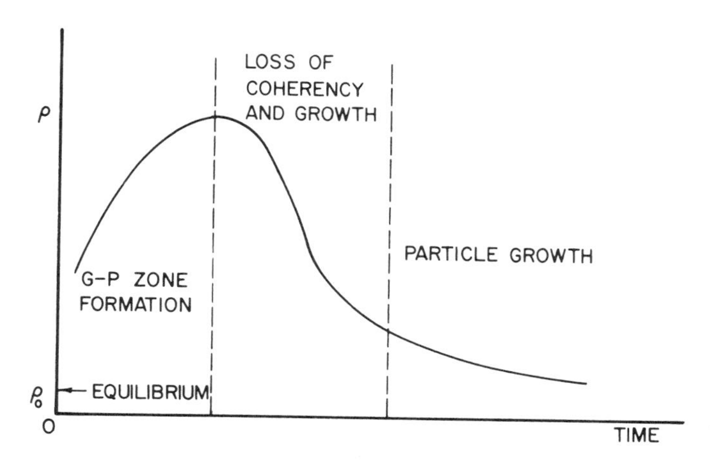

FIGURE 6-7. Schematic representation of the changes in electrical resistivity of an age-hardening alloy as a function of aging time at a given aging temperature.

6.5.2. Order-Disorder

The distortion effects of alloying ions in random solid solutions are dramatically illustrated by the near absence of these effects when ordering occurs within a given alloy. In this event, the alloy ions occupy preferred sites rather than random ones in the host lattice. Such a lattice, also known as a superlattice, has a much greater degree of spatial periodicity, or regularity, than the same alloy in the random state. The same is true for the periodicity of the potential within the lattice.

As noted at the end of Section 6.3, ordering of the alloy ions in the host lattice increases the regularity of the lattice and can result in a large decrease in resistivity because large decreases can occur in $\bar{k}(E_F)_T$. This is shown clearly in the "classic" Cu-Au system, in Figure 6-8. The quenched alloys, many of which are metastable, show the behavior previously noted in Figure 6-5c as being typical of random, substitutional solid solutions. When these alloys are aged and become ordered, large decreases in the electrical resistivity are observed. The maximum changes occur at 25 and 50% Au. Superlattices are formed. At the 25% alloy, nearly all of the Au atoms occupy corner sites in the FCC lattice. At the 50% alloy, the Au atoms may be considered to occupy body-centered sites, with Cu atoms at the corners of the unit cell. This actually is a CsCl-type lattice. Both arrays represent much higher degrees of lattice regularity and uniformity than those of random disordered, substitutional solid solutions. Under these conditions of long-range order, $\bar{k}(E_F)_T$ becomes smaller because the scattering by the more regular lattice is much less than that of a random solid solution. From Equation 6-27 it can be seen that the electrical resistivity would decrease. The total wave vector decreases because the decreased scattering results in longer effective electron wavelengths. Alloys of compositions other than the two discussed here have structures with varying degrees of order and disorder.

It also will be noted that any significant decreases in the resistivity caused by ordering will be accompanied by increases in the temperature coefficient of resistivity (Equation 6-31).

Alloys showing this metastable behavior are not normally used for electrical or electronic components.

6.5.3. Allotropic Changes

Many elemental metals, and their alloys, possess different crystal lattices known as

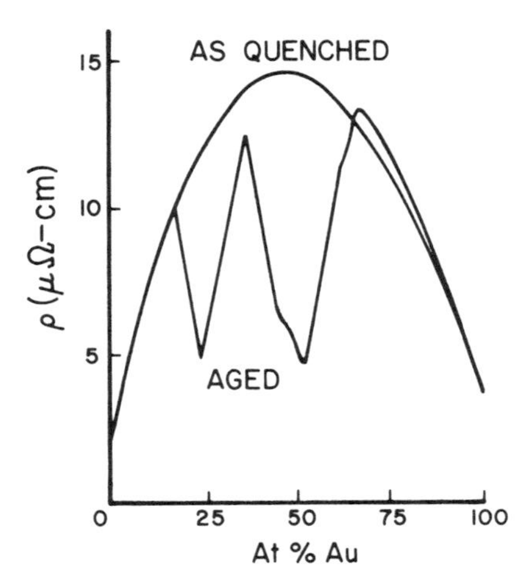

FIGURE 6-8. Effects of ordering upon the electrical resistivity of alloys of the Cu-Au system. (After Barrett, C. S., *Structure of Metals*, McGraw-Hill, New York, 1952, 288. With permission.)

allotropes. The lattice type present depends upon the ambient temperatures and pressures. Such transformations are phase changes. Each allotrope has different properties because each lattice is different. This results in different Brillouin zones for each allotrope.

When a metal or an alloy undergoes an allotropic change, $\bar{k}(E_F)_T$ changes because the lattice type changes. This is reflected in a change in the electrical resistivity. The resulting discontinuity of the curve of electrical resistivity as a function of temperature, thus, can be used to mark such a phase boundary.

Such measurements must be made using very slow heating and cooling rates. If this is not the case, nonequilibrium effects can mask the transition; the discontinuity may be suppressed. Hysteresis effects also can affect such determinations. For example, if an alloy contains elements with slow diffusion rates in the temperature range of interest, the transition could occur over a spread of temperatures. Upon cooling such alloys hysteresis effects also are observed. If the same slow absolute values of heating and cooling rates are employed, the allotropic change may be approximated as starting half way between the beginnings of the transformations noted upon heating and cooling.

A typical allotropic change as determined by electrical resistivity is shown in Figure 6-9.

6.6. EFFECTS OF DEFORMATION

The crystal lattices of cold worked metals are deformed and may be highly imperfect, depending upon the amount of deformation. They will remain so, to a large extent, unless thermal treatments are applied. The poor spatial periodicity of such lattices is very effective in scattering electrons. This also results in distorted Brillouin zones, accounting for the increased anisotropy of properties; these factors also cause large increases in $\bar{k}(E_F)_T$.

The working, or plastic deformation, of a metal or alloy below its recrystallization temperature, or below a phase reaction isotherm, results in a microstructure which consists of fine grains with orientations influenced by the kind, amount, and direction

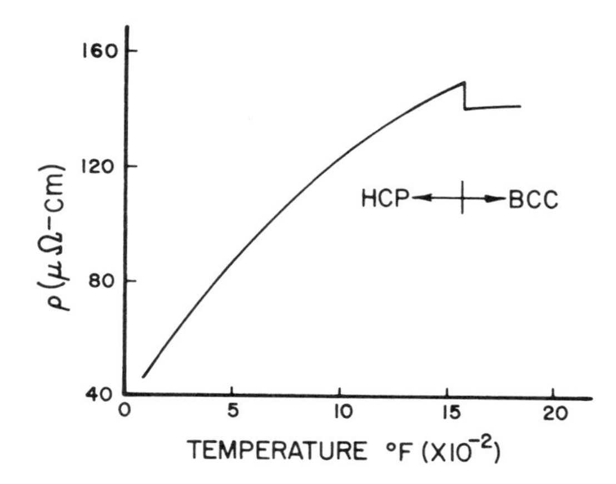

FIGURE 6-9. Allotropic change of titanium (99.9%) as detected by electrical resistivity. (From Lyman, T., Ed., *Metals Handbook, Properties and Selection,* Vol. 1, 8th ed., American Society for Metals, Metals Park, Ohio, 1961, 1225. With permission.)

of working. In addition, the structures of the small grains are highly imperfect and contain many types of lattice imperfections. Metals treated in this way would be expected to show even more highly anisotropic electrical properties because of the lattice distortions which result in "warped" Brillouin zones.

The electrical resistivity of metallic materials subjected to such treatments would be expected to show greater resistivities than those in the annealed condition. This is anticipated because $\bar{k}(E_F)_T$, Equation 6-14 or 6-27, would be expected to be increased by the condition of the highly imperfect distorted lattices of the grains.

The increase in resistivity induced by cold working is not as great as one might think. It appears that a maximum increase exists for a given metallic material. For example, the maximum increase in the resistivity of commercially pure metals lies between 4 and 8%. Very pure metals show increases of the order of up to 5%. Alloys show resistivity increases of from about 10 to 25%. The data indicate that purity greatly affects this behavior. It is known that the presence of alloying, or impure elements increases the recrystallization temperature. The purest material has the lowest recrystallization temperature. The mechanical energy which is converted to heat during the working process could be sufficient to recrystallize or to induce some degree of recovery or stress relief in the purer materials during the working process. The less pure, or alloy, materials could have been subject to some lesser degree of recovery or stress relief; these would show correspondingly higher resistivity increments, in agreement with the observed data. Such autogeneous thermal effects can obscure the true effects of cold working upon the electrical properties. Working suitably performed at cryogenic temperatures, with minimum energy inputs, can reveal virtually all of the effects of cold working.

Metals and alloys also show changes in electrical resistivity as they are strained within their "elastic" limits. The changes here are also a result of increases in $\bar{k}(E_F)_T$. This also is caused by lattice distortion. However, these changes are "elastic" and are considerably smaller than those observed in plastic deformation, since the lattice changes are smaller.

This property of metals is employed in strain gauges. Materials employed for such gauges must have a reasonably high elastic limit so that any measurable permanent deformation of the gauge, and erroneous results, are avoided. Such materials are selected on the basis of their gauge factor, γ, such that

$$\gamma = \frac{\Delta R/R}{\Delta L/L} = \frac{\Delta\rho}{\rho\epsilon} \tag{6-33}$$

where R is the resistance, L is the length of the gauge, and ε is the strain. Acceptable materials for such applications have $\gamma \simeq 2$ and can undergo strains up to 2%. In addition, such materials must possess small values of α (Equation 6-29) in order that errors introduced by temperature changes do not adversely affect the gauge readings. Materials commonly in use for this purpose are constantans (Cu-Ni) and manganins (Cu-Mn-Ni); both types of alloys have very flat, parabolic curves of temperature vs. resistivity in the neighborhood of room temperature (see Section 6.3 and Table 6-3). Their use largely eliminates errors in strain readings which may be introduced during testing by changes in temperature.

6.7. COMMERCIALLY AVAILABLE ALLOYS

Large numbers of alloys intended for electrical applications are commercially available. Many alloy producers provide very similar alloys, each under its own trade name. Representative alloys of the most frequently used classes are discussed here as representatives of the multiplicity of alloys in a given class.

The electrical properties of some of the large number of commercially available classes of alloys are given in Table 6-3. Alloys for electrical applications are melted and processed very carefully to maintain their compositions within narrow limits and to avoid impurities and contaminations. These may be compared to the corresponding properties given for the elements in Tables 5-6 (Chapter 5, Volume I) and 6-1. The alloy selected by the engineer will depend upon the specific application.

Some of the copper-nickel alloys have low resistivities. These are offered to provide ranges of resistivities, temperature coefficients, and sometimes to carry high currents. Here it becomes necessary to design the resistors to help dissipate the Joule heat. This heating effect must also be taken into account in the determination of the resistances of the components of the circuits at the temperatures of operation. Relatively large changes in resistance can occur when the temperature coefficient of resistance is large. These must be known in order that the components will be able to perform the desired functions in the circuit under operating conditions.

The commercial constantans, such as Advance, Cupron, and Copel, are used in resistors as well as in thermocouples (see Section 7.14.2, Chapter 7). Their use in resistors is based upon their medium resistivities and their relatively flat, parabolic resistance-temperature characteristics.

Many constantans have very flat, nearly parabolic resistance-temperature curves whose maxima are in the neighborhood of room temperature. These properties are a result of Brillouin zone overlap (see Section 6.3). The curve shapes and locations of the maxima are controlled by manipulating the compositions of this class of alloys. This family of alloys varies in composition from approximately 50% Cu − 50% Ni to 65% Cu − 35% Ni, each alloy within the range differing somewhat in properties. Most constantans contain small but significant quantities of Fe, Mn, Co, and other alloying elements, depending upon the source. These are almost invariably purchased in the annealed condition. It is customary to stress relieve components made from them after fabrication.

The small temperature coefficients of resistance, resulting from the flat resistance-temperature curves, make constantans desirable for use in components which will not change appreciably in resistance with temperature changes in the vicinity of room temperature. The large thermoelectric power of constantans vs. copper must be considered

Table 6-3
PROPERTIES OF SOME TYPICAL RESISTANCE ALLOYS

Trade name	Nominal composition (wt %)	Resistivity at 20°C $\mu\Omega$ − cm	Temperature coefficient $\Omega/\Omega/°C \times 10^3$ (near room temp.)	Thermoelectric power vs. cooper $\mu V/°C$ (near room temp.)
No. 30 Alloy	98 Cu, 2 Ni	6	+ 1.5	− 13.9
Lohm	94 Cu, 6 Ni	10	+ 0.8	− 14.3
No. 90 Alloy	88 Cu, 12 Ni	15	+ 0.4	− 24.5
No. 180 Alloy	78 Cu, 22 Ni	30	+ 1.8	− 35.4
Midohm	77 Cu, 23 Ni	30	+ 0.3	− 37.0
Advance	57 Cu, 43 Ni	49	± 0.04	− 43.0
Cupron	55 Cu, 43 Ni	49	± 0.04	− 42.4
Copel	55 Cu, 45 Ni	49	± 0.04	− 42.4
Manganin	87 Cu, 13 Mn	48	± 0.02	+ 0.9
Manganin	84 Cu, 12 Mn, 4 Ni	48	± 0.01	− 1.5
Shunt Manganin	83 Cu, 11 Mn, 4 Ni	45	± 0.01	− 1.8
ISA	82 Cu, 12 Mn, 5 Ni	45	± 0.02	− 2.0
NCM	70 Cu, 10 Mn, 20 Ni	45	± 0.02	− 10.0
Isotan	55 Cu, 1 Mn, 45 Ni	45	± 0.04	− 40.0
A Nickel	99 Ni	10	+ 5.0	− 22.0
Tophet A	80 Ni, 20 Cr	110	+ 0.1	+ 4.8
Chromel A	80 Ni, 20 Cr	110	+ 0.12	+ 4.8
Nikrothal - L	75 Ni, 17 Cr + Si + Mn	134	± 0.02	− 1.0
Evanohm	75 Ni, 20 Cr + Al + Cu	134	± 0.02	+ 0.8
Karma	73 Ni, 20 Cr + Al + Fe	134	± 0.03	+ 3.0
Balco	70 Ni, 30 Fe	20	+ 4.5	− 39.9
Hytemco	72 Ni, 28 Fe	20	+ 5.0	− 39.2
Tophet C	61 Ni, 24 Fe, 15 Cr	113	+ 0.13	− 0.8
Nichrome	60 Ni, 24 Fe, 16 Cr	113	+ 0.2	− 0.8
Nichrome V	80 Ni, 20 Cr	110	+ 0.1	+ 4.8
Chromel C	60 Ni, 24 Fe, 16 Cr	110	+ 0.2	+ 0.8
Comet	65 Fe, 30 Ni, 5 Cr	95	+ 0.6	− 3.9
Chromel D	47 Fe, 35 Ni, 18 Cr	100	+ 0.3	− 2.6
Chromax	45 Fe, 35 Ni, 20 Cr	100	+ 0.4	− 2.6

From Lyman, T., Ed., *Metals Handbook, Properties and Selection*, Vol. 1, 8th ed., American Society for Metals, Metals Park, Ohio 1961, 780. With permission.

in applying these alloys to components for very accurate instrumentation. The constantans can form small thermocouples with the copper connecting wires. These can create small parasitic voltages and currents when temperature differentials exist throughout the electrical circuitry. They can result in the introduction of errors in high-accuracy devices intended to measure small voltages or currents.

Manganin, like constantan, is a generic term for a class of alloys for resistors. Manganin-type alloys usually are solid solutions of copper, manganese, and nickel. Many manganin-type alloys are characterized by very flat nearly parabolic resistance-temperature curves whose maxima are in the neighborhood of room temperature. The temperature coefficients of many manganins over a narrow temperature range can be smaller than those of constantans. Manganins have another advantage over constantans in that their thermoelectric powers vs. copper are quite small. This minimizes any extraneous sources of error. Manganins also demonstrate a very high degree of electrical stability when properly stress relieved and protected from strains (see Section 6.6) and injurious atmospheres.

The maximum point (peak temperature) and curve shape of the parabolic resistance-temperature relationship of manganin also are a result of zone overlap and are controlled by variations in alloy composition. This is desired so that the effects of any change in the ambient temperature of the instrumentation will be minimized and its calibration will remain unaffected. The temperature coefficient of commercial manganin is generally less than $\pm 0.00001\ \Omega/\Omega/^\circ C$ for a 10°C interval on either side of the peak; this is usually selected to be near the ambient temperature of operation of the instrument. When instrumentation is designed to operate at higher temperatures, the chemical analysis of the alloy is changed to shift the peak temperature to the desired value. So-called shunt manganin, which carries high currents, and consequently heats considerably in use, usually has its peak temperature in the range from 45 to 65°C, depending upon the amount of heating to be generated.

The above combinations of properties have led to the almost universal use of manganin in the construction of very high-accuracy resistors, slidewires (potentiometers), and other components for electrical measuring and control devices. Manganin alloys are always purchased in the annealed condition. The components made from them are stress relieved after fabrication.

The Ni-Cr-Al base alloys (Evanohm and Karma) have resistivities of more than twice that of manganins and nearly parabolic resistance-temperature curves. This means that accurate resistors of higher resistance values can be made to occupy small volumes. Their temperature coefficients of resistance are less than $\pm 0.00002\ \Omega/\Omega/^\circ C$ over the range from −50° to +100°C. The average thermoelectric powers vs. copper of this class of alloys is small. These properties are excellent for accurate resistors. The electrical stability of this class of alloys can be very sensitive to mechanical and thermal treatments. They also have the drawback that corrosive fluxes are required to produce acceptable soldered or brazed joints. If such fluxes are not completely removed, they continue to corrode and cause electrical instability. Components made from these alloys also are stress relieved after fabrication.

The Ni-Fe-Cr and the Fe-Ni-Cr alloys are not generally used for accurate resistor applications. These alloys have relatively high temperature coefficients of resistance. They are used most widely as electric heating elements.

Many of these alloys, especially those containing large amounts of nickel and chromium, also are used for high-temperature, heat-resistant components in oxidizing atmospheres. Typical applications of this kind would be their use as furnace components. Such applications should be made with caution; some of these alloys deteriorate in reducing atmospheres and in those containing sulfur, even in very small amounts. This susceptibility also holds when these alloys are used as heating elements.

6.8. SUPERCONDUCTIVITY

The behavior of the electrical resistance of many normal metals and alloys is described earlier in this chapter. These materials decrease in resistivity with decreasing temperature and approach their residual resistivity as T approaches 0 K. Certain other materials do not show this behavior; their resistances more or less abruptly drop to zero at various temperatures below about 24 K, depending upon the material. This means that their electric fields are zero in the superconducting state. This behavior was first discovered by H. Kamerlingh Onnes (1911) while working with mercury. Such behavior is shown schematically in Figure 6-10.

The absence of resistance in the material enables currents to flow for long times. Under certain conditions the decay time of the supercurrent can be the order of 10^5

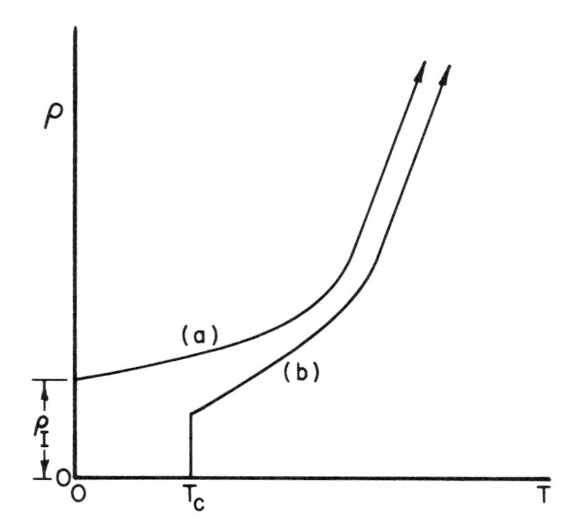

FIGURE 6-10. Electrical resistivity at cryogenic temperatures. (a) Normal metal and (b) Type I superconductor.

years. Shorter decay times are observed in other superconducting magnetic materials because of irreversible magnetic flux effects within the superconductor; these are described subsequently.

The temperature at which the resistance goes to zero, in the absence of a magnetic field, is the transition temperature, T_c. Many metallic elements, alloys, intermetallic compounds, and semiconductors show this behavior. Materials which do not show superconducting properties are designated as being "normal" in the sense that they show residual resistivities.

The transition temperatures of some elements are given in Table 6-4. The resistivities of these elements tend to be high at normal temperatures, indicating strong electron-ion interactions. Superconducting elements are infrequent, or not present, in groups IA, IB, 2A, 3A, 5B, 6B, 7B, and 8B of the Periodic Table. Many of the elements in the remaining groups (2B, 3B, 4A, 4B, 5A, 6A, and 7A) are superconductors. The latter set has relatively high resistivities arising from numerous electron-phonon interactions (see Table 6-4). This is important in the superconductivity mechanism.

It has been demonstrated empirically that the average number of valence electrons per atom must lie between two and eight for superconductivity to occur; a complex relationship has been used to approximate the effect of the electron-ion ratio on the transition temperature. The transition temperatures of alloys of superconductors with superconductors, or of superconductors with normal metals, may be higher or lower than those of the superconductors involved. Some of these alloys can show superconducting transitions which occur over a range of temperatures, rather than at a specific temperature. This range can be affected by such factors as prior metallurgical history, stress, and purity.

Many intermetallic compounds show superconductivity. Some of those with the highest values of T_c are given in Table 6-5. Most of these contain at least one constituent which is a superconductor. In the small number of compounds which do not contain an elemental superconductor, one of the constituents occupies a place in the Periodic Table next to a superconductor. It also appears that the role of the lattice type is important. The sodium chloride, nickel arsenide, β-tungsten, and σ-phase structures are found frequently.

Table 6-4
SUPERCONDUCTING PROPERTIES OF SOME SELECTED SUPERCONDUCTING ELEMENTS

Element	T_c(K)	H_o(Oe)	Element	T_c(K)	H_o(Oe)
W	0.012	1070	Th(α)	1.37	162
Be	0.026	—	Pa	1.4	—
Ir	0.14	19	Re	1.7	193
Hf(α)	0.165	—	Tl	2.4	171
Ti(α)	0.39	56	In	3.4	293
Ru	0.49	66	Sn(β)	3.7	309
Cd	0.52	30	Hg	4.15	412
Os	0.65	65	Ta	4.48	830
U(α)	0.68(?)	—	V	5.3	1020
Zr(α)	0.55	47	La(β)	5.9	1600
Zn	0.85	52	Pb	7.2	803
Mo	0.92	98	Tc	8.2	1410
Ga	1.09	59	Nb	9.2—9.4	1950
Al	1.19	99			

From Savitskii, E. M., *Superconducting Materials*, Plenum Press, New York, 1973, 82. With permission.

Table 6-5
SUPERCONDUCTIVITY OF SELECTED COMPOUNDS

Compound	T_c (K)
Nb_3Sn	18.05
Nb_3Ge	23.2
Nb_3Al	17.5
NbN	16.0
$(SN)_x$ polymer	0.26
V_3Ga	16.5
V_3Si	17.1
$Pb_1Mo_{5.1}S_6$	14.4
Ti_2Co	3.44
La_3In	10.4

From Kittel, C., *Introduction to Solid State Physics*, 5th ed., 1976, 338. With permission.

When a metal becomes superconducting no change occurs in its crystal structure, nor does recrystallization take place. In the absence of a magnetic field no latent heat is evolved. However, very small changes in lattice parameter and volume do occur. The lattice becomes slightly more perfect. These changes are of the order of one part in 10^7. The molar entropy change resulting from this is about $10^{-3}Nk_B$. Observable changes in elastic properties and thermal expansion have not been noted, despite theoretical predictions that these properties should be very slightly affected. No Thomson heat is generated below T_c. This indicates that no electron entropy changes occur (see Sections 7.4 and 7.5, in Chapter 7).

Experiments have shown that superconductivity disappears upon the application of a critical magnetic field, H_c. An approximation of H_c, for long, cylindrically shaped specimens, is given by

$$H_c = H_o \left[1 - \frac{T^2}{T_c^2}\right] \tag{6-34}$$

in which H_o is the critical field at 0 K and T_c is the temperature at which the material becomes superconducting in the absence of a field. This behavior is shown schematically in Figure 6-11 along with a relation of H_o to T_c.

Onnes (1913) showed that an electric current in a wire would eliminate its superconductivity if the current exceeded a certain amount. It was shown later, by Silsbee (1916), that the field induced by the current, rather than the current itself, was responsible for this, as discussed below.

Not only does the electrical resistance go to zero, but a superconductor also is virtually a perfect diamagnet. It is shown, in Chapter 8, that when the magnetic flux is changed, a current is induced in the direction opposing the flux change (Lenz's law). The electrons induce a magnetic moment in opposition to the applied field. All superconductors react like nearly perfect diamagnets and reject virtually all of the flux from their interiors. Under certain conditions, some may show mixed behavior and permit flux penetration.

One of the important properties of superconductors is the average depth of penetration, λ, of the field into the bulk material. It was found, on a semi-empirical basis, that λ increases with temperature. This was based upon the idea that the "number of electrons" entering into this process decreases with temperature. The relationship between λ and T is

$$\left[\frac{\lambda(0)}{\lambda(T)}\right]^2 = \left\{\left[\frac{T}{T_c}\right]^4\right\}^{-1} \tag{6-35}$$

in which $\lambda(0)$ and $\lambda(T)$ are the penetrations at 0 K and at $T < T_c$, respectively. This relationship has wide use in penetration analyses. In a normal dielectric metal λ is about 500 Å at 0 K, while that of a superconductor is less than 200 Å.

Thus, under superconducting conditions, the magnetic induction, B, inside the superconductor is zero. Virtually all of the lines of flux are external to the bulk of the material. This is known as the Meissner effect and is shown in Figure 6-12.

Since $B = 0$, the magnetization is $-M = H$ (Equation 8-13); its permeability is zero (Equation 8-14); and its magnetic susceptibility $\chi = -1$ (Equation 8-9). Thus, the superconductor is perfectly diamagnetic, except for the very small extent of field penetration. Materials reacting in this way are called Type I, or "ideal", or "soft", superconductors (See Figure 6-12).

It will be noted that the field intensity is greatest at A-A′ in Figure 6-12. An increase in the field can cause H_c to be reached in the volumes of the material adjacent to the concentrated field. These volumes, therefore, can revert to the normal state and the remainder of the bulk material stays superconductive. A mixture of regions may exist under these circumstances. The extent of the normal material increases as the field strength increases beyond H_c (Figure 6-13). Transition to the completely normal state occurs when the upper critical field, H_{c2}, is reached. Materials showing this kind of behavior are called Type II, or "hard", or "high-field", superconductors.

The magnitude of the upper critical field, H_{c2}, may be of the order of 100 times the value of H_c. Solenoids made of hard superconductors have given fields greater than 100kG.

Qualitative descriptions of the mechanisms responsible for these phenomena are given because the complexity of their mathematical descriptions is beyond the scope

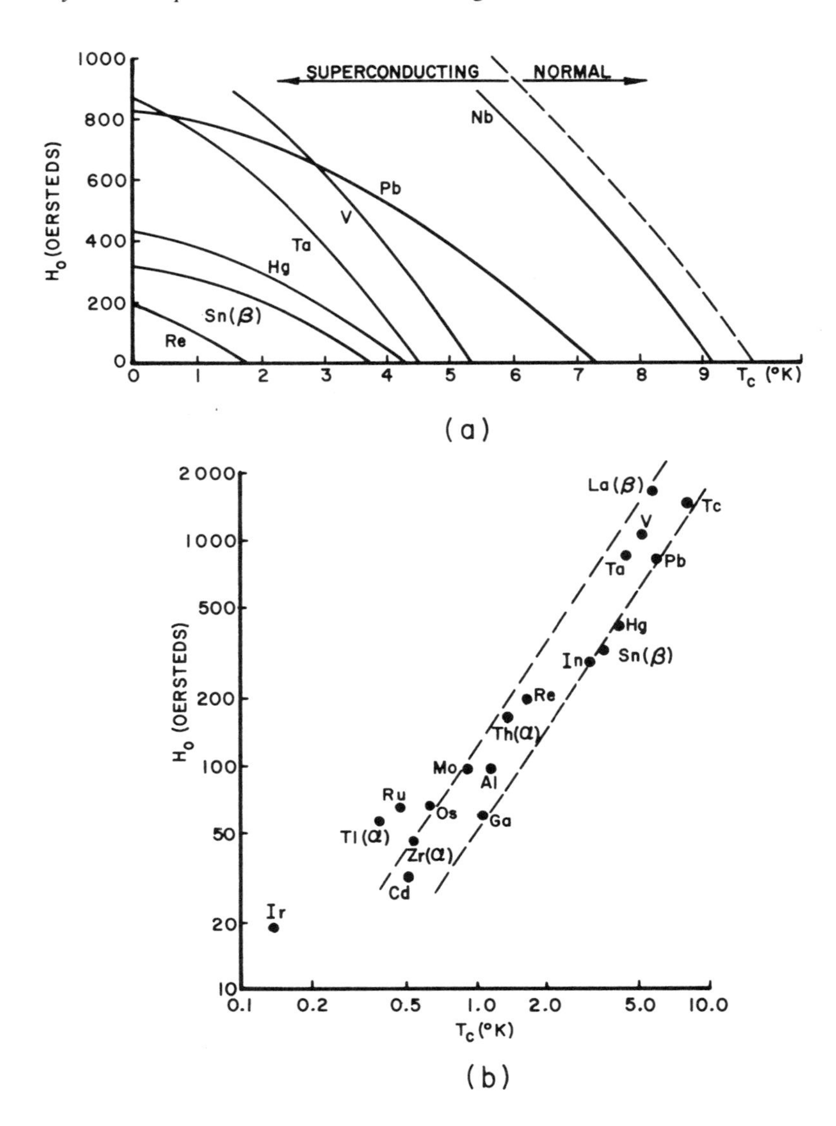

FIGURE 6-11. (a) Schematic diagram showing onset of superconductivity along with the behavior of several elements; (b) log H_o as a function of log T_c. Note that most of the elements lie within a narrow band. See Table 6-4. (From Savitskii, E. M., et al., *Superconducting Materials*, Plenum Press, New York, 1973, 82. With permission.)

of this text. Some of the early ideas are presented to lead up to and help clarify the modern theory.

One early model, known as the "two-fluid theory", assumed that part of the electrons were in the ground state and were superfluid. The remaining electrons were considered to behave normally. This approach was able to provide explanations for the electromagnetic properties, diamagnetism, and zero resistivity.

F. London (1935) first suggested that electron interactions involving electron coupling and an energy gap were fundamental to the understanding of superconductivity. The energy gap was postulated to separate the electrons in the lowest states from those in higher, excited, levels. This model also considered that a high degree of uniformity of the momentum of electrons was required. This resulted from the finding that "the

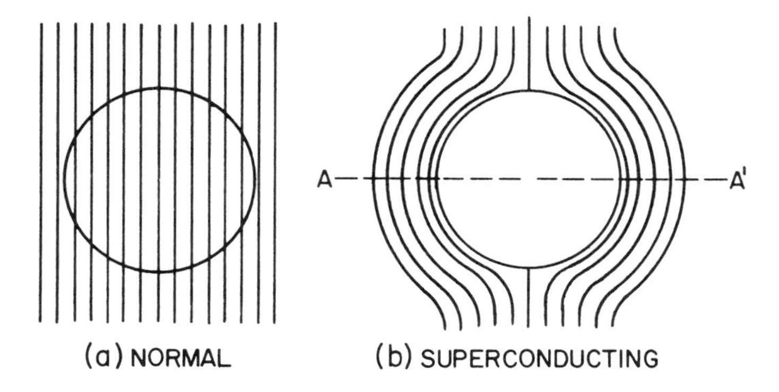

FIGURE 6-12. Flux lines in a transverse field. (a) about a normal metal and (b) the exclusion of flux by a superconductor known as the Meissner effect.

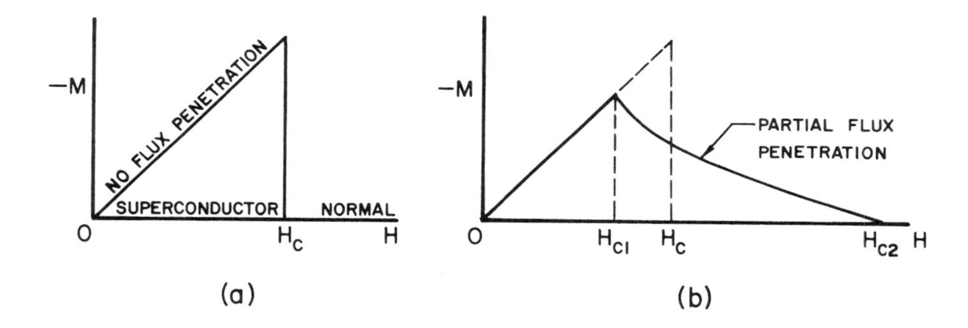

FIGURE 6-13. Magnetization curves for (a) a Type I superconductor and (b) a Type II superconductor.

wide extension in space of the wave functions representing the same momentum distribution throughout the whole metal in the presence as well as the absence of a magnetic field'' requires that ''the long-range order of the (average) momentum is to be considered one of the fundamental properties of the superconducting state''. In effect, this is a macroscopic model based upon a ''kind of solidification or condensation of the average momentum distribution''. These ideas are the bases for the present theory in which paired ''electrons'', all with the same momenta, account for the ''solidity'', or rigidity, of the distribution of momentum; all of the paired particles must be considered to have the same momentum, as described below.

Other investigators had shown T_c to be an inverse function of the square root of the isotopic mass of Hg. This led J. Bardeen and others to the idea that superconductivity might result from electron interactions with zero-point vibrations because lattice frequencies show a similar dependency. (This can be seen in an elementary way by substituting Equation 4-40 into Equation 4-113 and obtaining θ_D as a function of $M^{-1/2}$.) It was also thought that an energy gap might result from such electron-phonon interactions. This may be illustrated in an oversimplified way by considering the electrons in a lattice. An ion in the lattice becomes polarized when an electron interacts with it. Small volumes of the lattice surrounding it become distorted momentarily. Another electron of opposite spin, and with energy close to that of the first electron, is affected by this local phonon; it is attracted to it and simultaneously repels other electrons. The reaction between the first electron with the ion lowers the energy of the second electron which is attracted to the positive charge on the ion. This results in an attraction (or less repulsion) between the two electrons; their mutual electrostatic repulsion is diminished to a degree which permits electron-pair formation. It now is considered that quasi-particles (Section 5.7) condense into bound pairs in a *single* state just below the uppermost, normally filled level at 0 K. The behavior at temperatures below T_c is controlled by the formation of normal quasi-particles (Section 5.7) at levels above that of the bound pairs. This results in an energy gap between the active normal states and that of the bound pairs.

Field theory analyses by Frölich (1950) showed that mutual reactions between electrons and phonons can result in attractive interactions among pairs of electrons near E_F. The optimum conditions for this to occur are that the two electrons have opposite spin and momenta and be separated by less than about 1 Å, the coherence length. This led to models based on the treatment of electron-pairs as pseudo-molecules. Bardeen suggested that if the application of the concept of bound pairs was applied to quasi-particles, it could be an important factor in superconductivity theory.

It was later shown by L. N. Cooper (1956) that if there is such an attractive interaction, a pair of quasi-particles will form a bound state, even for weak attractions. Such pairs now are known as Cooper pairs. The Fermi level, stable for normal metals, was shown to be unstable with respect to the formation of such bound pairs in superconductors. Calculations showed that the bound pairs of quasi-particles formed from all of the quasi-electrons within the range of about k_BT_c of E_F would overlap because the distance between such pairs is much less than the size of a pair and results in a high degree of pair overlap. Thus, the pairs cannot be considered to show independent motions because of this overlap. This showed the pseudo-molecule models to be incorrect, but that the London concept of paired particles, all with the same momentum, was reasonable.

Normal metals can be described by the way in which the quasi-particles occupy states in $\bar{k}$ space near or above E_F (see Section 5.7). But, as previously noted, the Fermi level is not stable with respect to Cooper pairs for $0 \text{ K} < T < T_c$. However, the energy of a quasi-particle can depend upon the distribution of other quasi-particles. This causes the solution of the Hamiltonian for the configurations of paired quasi-particles to be inexact. They are sufficiently accurate, however, when the excitation energies are not excessive. The resulting paired particles are assumed to include quasi-particle self-energies and correlation energies. The lowering of the energy of each pair is equal to the bonding energy of the pair. A component of this also arises from the simultaneous interaction of large numbers of pairs; this causes an additional attraction between members of each pair. The superconducting properties were thought to result from the residual attractions among the normal quasi-particles with the paired ones. These now are considered to result from a dynamic equilibrium between both types of particles.

The original work by Bardeen, Cooper, and J. S. Schrieffer (1957), also known as the BCS theory, employed a wave function for the ground state of a superconductor which was a linear combination of normal quasi-particle configurations where the quasi-particle states were filled by pairs of opposite spin; all such pairs have the same total momentum and zero net spin. Exclusion is invoked here to require that both states of a pair be filled by pairs, or that both be unfilled. All such pairs have the same total momentum; the sum of the wave vectors of a given pair is the same as those of all other pairs. For current to flow the wave vector sum of any given pair cannot be equal to zero. Such a pair, with total spin equal to zero, is a boson; thus, many such particles can occupy a single state and have the same momentum (see Section 5.4.1, Chapter 5, Volume I).

The energy gap, E_g, unlike that of a semiconductor, is not a function of E_F; it is temperature dependent and has no definite location in $\bar{k}$ space; E_g decreases to zero at T_c. Both of these properties, E_g and E_F, of a superconductor are shifted in $\bar{k}$ space and momentum space when current flows at different temperatures below T_c, in a manner analogous to that given by Equation 5-58. This results from the dynamic equilibrium between the paired and unpaired particles. Excitations of the paired, superfluid, quasi-particles result in a current. However, when a local equilibrium is established which closely corresponds to that at 0 K, a net flow of the paired quasi-particles exists; this is equal to $\varrho_s v$, where ϱ_s corresponds to the density of the superfluid in the two-fluid model (the density of the paired quasi-particles) and v is the rate of flow. As $T \rightarrow T_c$, $E_g \rightarrow 0$ and $\varrho_s \rightarrow 0$. This results from the thermally induced splitting up of the paired particles and not from scattering, nor from an "evaporation" of pairs. Below T_c the current continues to flow, despite scattering, because scattering does not affect the momentum of any of the other Cooper pairs. At a $T < T_c$, excitations of the quasi-particles reduce the current until an equilibrium is reached and a new value of v is established. Increasing T causes increasing quasi-particle excitations until, at T_c, E_g, and ϱ_s vanish and normal behavior returns. (It should be noted that a group of superconductors exists which seems not to have a gap.)

The BCS theory found that T_c is a function of the density of states within about k_BT_c of E_F, $N(E_F)$, and the particle-lattice-pair reaction coefficient, A. This is given by

$$T_c = 1.14\,\theta_D \exp\left[-1/AN(E_F)\right]$$
$$AN(E_F) \ll 1 \qquad (6\text{-}36)$$

As the temperature decreases to T_c, quasi-particles in the range $\pm\, k_BT_c$ of E_F form Cooper pairs; an abrupt transition to superconduction occurs. This results in $E_g \cong 2k_BT_c$, an energy of about 10^3 to 10^4 times that of a single Cooper pair. At 0 K all of the pairs are in the superconducting ground state. Since these are bosons, they occupy a single state with an energy approximately k_BT_c below the uppermost filled normal state. The normal particles must occupy individual states singly, as required by the principle of exclusion. For $0\ K < T < T_c$ a dynamic equilibrium exists, for a given T, between the normal and paired particles; E_g must, therefore, change (see Figure 6-14). The presence of the factors in Equation 6-34 now may be considered on an intuitive basis. The higher θ_D is, the more uniform the lattice vibrations will be at very low temperatures. This gives a greater probability for the particle-lattice-pair reaction to take place. This leads to the particle-lattice-pair reaction coefficient, A, which also must be taken into account; without pair production superconductivity could not occur. The number of particles able to participate in pair production, $N(E_F)$, obviously must be considered. This is approximately 10^{-4} of those available for transport processes in normal metals at ordinary temperatures.

When current flows, each Cooper pair moves in the same direction as the current with the same velocity as all other pairs. Some of the pairs may split up, as a result of thermal excitation, and the individual particles may jump the gap to fill empty higher energy states. The other Cooper pairs remain together and resist separation. All pairs have the same momentum when an electric field is applied; all have the same drift velocity. This flow constitutes the superconducting current. Since all pairs must have the same momentum, the scattering of some Cooper pairs cannot change the momentum of any of the other pairs. The momentum of all other pairs remains unchanged and the current continues to flow. Thus, the scattering is ineffective, as shown below.

Because the density of the Cooper pairs is high, only a small drift velocity is required

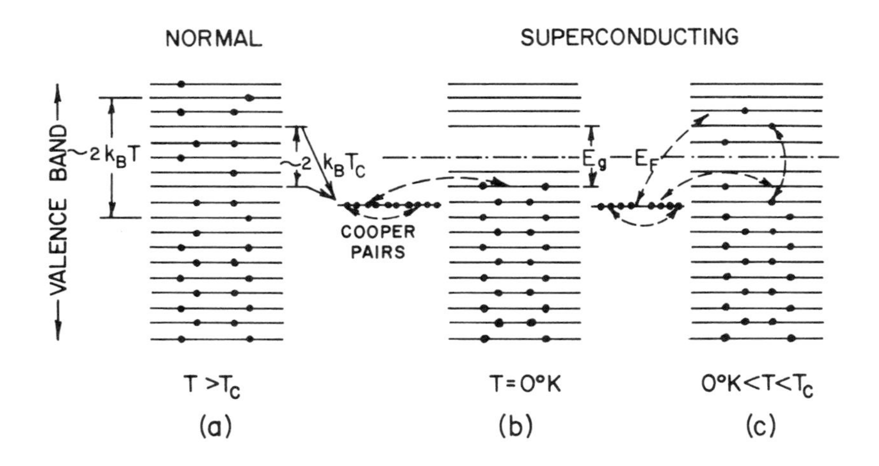

FIGURE 6-14. Schematic diagram of the transition from (a) the normal state; (b) to the superconducting state at 0 K and the formation of Cooper pairs which creates an energy gap in the valence band; (c) the dynamic equilibrium in the production of Cooper pairs at a given temperature in the range of temperatures below T_c; note that the Cooper pairs, being bosons, occupy a single energy state. (After Hurd, C. M., *Electrons in Metals*, John Wiley & Sons, New York, 1975, 207. With permission.)

to generate an appreciable current. The small momentum required for this corresponds to a large de Broglie wavelength. In general, significant amounts of scattering will occur when the wavelength is about the same, or smaller, than the dimension of the scattering source. The large wavelengths in this case help to explain the ineffectuality of scattering and the persistance of superconducting currents; they are the reason for the small amount of scattering.

Continuous scattering of Cooper pairs does occur. However, the number of pairs at any T, for $0 \text{ K} < T < T_c$, as discussed previously, is a result of the dynamic equilibrium between the normal quasi-particles and the Cooper pairs, and their number is maintained. The equilibrium density of the pairs can enter into a resistanceless flow. When a voltage is applied to a superconductor, both the pairs and the normal quasi-particles transport the current. The Cooper pairs behave as though they short-circuit that part of the current carried by the normal particles and produce the resistanceless, superconducting current.

The phonons which generate the Cooper pairs have a maximum velocity equal to the speed of sound in the crystal. If the pairs are made to travel through the lattice at velocities approaching this limiting value, further pair production cannot proceed. The equilibrium in pair production is upset and the number of pairs decreases to zero.

The depth of penetration, λ, of the magnetic field into a superconductor, in the Meissner effect, is approximately the same as the coherence length. (It will be recalled that the coherence length is that distance between two quasi-electrons of opposite spin and equal and opposite momenta most favorable for the formation of a Cooper pair.) This varies between 10^{-7} and 10^{-6} m. The penetration is, as would be expected, a function of the material and the temperature. The dependence of λ upon the coherence length explains the two types of superconductors in Figure 6-13. Those with Type I characteristics have coherence lengths larger than λ; they show uncomplicated magnetization curves because of this. Type II superconductors result when the coherence length is less than λ. This accounts for the mixed behavior between the onset of superconductivity and the lower critical value of the applied magnetic field. The lower the field that is applied to such materials, the smaller the volume within their bulk which will remain in the normal state, until, at the lower critical field, the entire volume is superconducting.

The reversion to the normal state induced by the application of a magnetic field is a result of splitting of the Cooper pairs by the field. Each particle in a pair is subject to a force which is proportional to the magnetic flux density and which is perpendicular to the motion of the particle. This induces a curvature in the path of the particle without changing its energy; it is known as the Lorentz force. When this force becomes sufficiently large, it can break up the Cooper pairs and account for the properties shown in Figures 6-11 and 6-13.

Several major problems still remain to be explained by the theory. The reasons why some elements are superconducting while others are not have not been given as yet. The mechanisms responsible for specific superconducting alloys and compounds also are still awaiting description. Thus, it is not possible, at present, to predict which materials will be superconductors. The search for materials with higher transition temperatures, therefore, is essentially empirical at present; it is based upon the general behavior of the classes of materials which were described earlier in this section.

Some of the more obvious applications of superconducting materials include their applications in very high-field magnets (one of which was described previously), and in the transmission of electric power. The former are commercially available, and are used primarily for research applications. The problem in their application for power transmission resides in their very low values of T_c. Extensive cryogenic refrigeration equipment is required for this, especially for power lines of significant length. The costs for this are prohibitive at the present time. In addition, the magnetic field set up by the current would eliminate the superconducting state. Other applications, such as the use of superconducting materials as components in electronic circuits, including computers, also are affected adversely by the costs of refrigeration.

6.9. PROBLEMS

1. The electrical resistivity of very pure Pt can be approximated by $R(t) = R(0)\,[1 + 3.9788 \times 10^{-3}\, t]$ where R(0) is given by 9.83 $\mu\Omega$−cm. Explain the difference between this material and that cited in Table 6-1.
2. Discuss the reasons for the change in slope of the curve of the relative resistance of Ni vs. temperature above and below 400°C (see Reference 12, p. 1218).
3. Calculate the electrical resistivities and temperature coefficients of resistivity of silver-base binary alloys containing 1 wt% Zn and Ge, respectively. Correct values to 25°C.
4. Calculate the electrical resistivities and temperature coefficients of resistivity of silver-base alloys containing 5 wt% of Cd and In, respectively. Correct values to 25°C.
5. Calculate the Matthiessen constants for Cu-, Ag-, and Au-base alloys. Explain any significant differences.
6. Calculate the electrical resistivity and temperature coefficient of alloys containing 87 Cu, 13 Mn and 83 Cu, 13 Mn, 4 Ni (wt%), respectively.
7. Draw schematic curves for the electrical resistivities of the alloys of the Ag-Ti system for the 700°C isotherm.[11]
8. Draw schematic curves for the electrical resistivities and temperature coefficients of alloys of the Al-Ca system at 450°C.[11]
9. Draw schematic curves for the electrical resistivities of the alloys of the Sb-As system at the 375°C isotherm.[11]
10. Draw schematic curves for the electrical resistivities and temperature coefficients for alloys of the Au-Fe system at the 350°C isotherm.[11]

11. Make a sketch of the variation of resistivity for a given, severely cold worked solid solution alloy as a function of temperature up to recrystallization. Label areas with mechanisms responsible for the responses noted.
12. Discuss the reasons for the necessity for the use of slow heating and cooling rates when electrical measurements are employed to detect phase changes in the solid state.
13. Describe and discuss an electrical technique suitable for the detection of solid-state phase changes.
14. Discuss the reasons why precipitation-hardening alloys normally are not used for high-accuracy components for electrical circuitry.
15. Would an alloy capable of long-range ordering be used for accurate electrical circuitry? Give your reasons.
16. Describe some ways in which superconductors could be used in electronic calculators.
17. Discuss the major factors which must be taken into account in the design of superconducting power-transmission lines.

6.10. REFERENCES

1. Matthiessen, H. and Vogt, C., *Pogg. Ann.*, 122, S19, 1864.
2. Linde, J. O., *Ann. Physik*, 10, 52, 1931; 14, 353, 1932; 15, 219, 1932.
3. Robinson, A. T. and Dorn, J. E., *Trans. TMS-AIME*, 191, 457, 1951.
4. Hansen, M., Johnson, W. R., and Parks, J. M., *J. Metals*, 3(12), 1184, 1951.
5. Pollock, D. D., *Trans. TMS-AIME*, 230, 753, 1964.
6. Stanley, J. K., *Electrical and Magnetic Properties of Metals*, American Society for Metals, Metals Park, Ohio, 1963.
7. Savitskii, E. M., Baron, V. V., Vefimov, Yu., Bychkova, Mi., and Myzenkova, L. F., *Superconducting Materials*, Plenum Press, New York, 1973.
8. Bardeen, J., *Physics Today*, American Physical Society, 1963, 19.
9. Hurd, C. M., *Electrons in Metals*, John Wiley & Sons, New York, 1975.
10. Kittel, C., *Introduction to Solid State Physics*, 3rd. ed., John Wiley & Sons, New York, 1966.
11. Hansen, M. and Anderko, K., *Constitution of Binary Alloys*, McGraw-Hill, New York, 1958.
12. Lyman, T., Ed., *Metals Handbook*, Vol. 1, American Society for Metals, Metals Park, Ohio, 1961.

Chapter 7

THERMOELECTRIC PROPERTIES OF METALS AND ALLOYS

Thermocouples have been used extensively for well over 100 years as devices for the measurement of temperature and for process control. Such applications range from cryogenic temperatures up to about 2800°C (5072°F). They are employed in almost every industrial process, and in many research applications which involve the measurement and/or the control of temperature. They constitute accurate, relatively inexpensive means for accomplishing these purposes when they are properly used.

Under certain conditions thermocouples will show large deviations compared to their original calibrations. Many technical and scientific personnel, unaware of this, assume that thermocouples always give invariant results. Ignorance of the origin and nature of thermoelectric properties, and of the factors which affect them, continues to be a major factor in the production of much out-of-specification processed materials, spoiled materials, unnecessary expenditure of thermal energy, and research data which are either misleading or useless. Such costly errors make it imperative that technical personnel understand the limitations under which thermocouples may be used with a high degree of reliability.

In addition to their practical applications, suitably constituted thermocouples can reveal changes in the solid state both accurately and economically. The thermodynamic basis for thermoelectricity has been understood for well over 100 years. However, this does not provide mechanisms explaining thermoelectricity. It has been only since the advent of quantum mechanics that the understandings of some of the mechanisms have been available. These now provide the bases for greater insight into solid-state reactions, including phase equilibria, as well as the reasons why so few thermocouples are in common use; they also explain the limitations under which thermocouples may be used to provide reliable temperature measurements.

7.1. SEEBECK EFFECT

A thermocouple is a device used primarily for the measurement of temperature. Its operation is based upon the findings of Seebeck (1821). This work showed that a small electric current will flow in a closed circuit composed of two dissimilar metallic conductors when their junctions are kept at different temperatures (Figure 7-1). The electromotive force, emf, produced under these conditions is known as the Seebeck emf. The pair of conductors, or thermoelements, which constitutes the thermoelectric circuit is called a thermocouple. Conductor A is defined as being positive to B if the current flows from A to B at the cooler junction. Simply stated, a thermocouple is a device which converts thermal energy into electrical energy. The amount of electrical energy produced can be used to measure temperature.

Certain pairs of thermoelements give emfs which vary in a regular way with the temperature differences between their junctions. Thermometric use is made of this behavior by maintaining one of the junctions at a known, fixed, reproducible temperature. This temperature is called the reference temperature, and is usually taken as the melting point of ice (0°C) for practical measurements. The junction maintained at such a temperature is known as the reference junction; the other junction, held at the temperature to be determined, is called the measuring junction. When a reference temperature is used, the temperature difference, and consequently the emf of a thermocouple, becomes a function of the temperature of the measuring junction. This permits the establishment of tables, curves, or mathematical relationships which provide the emfs

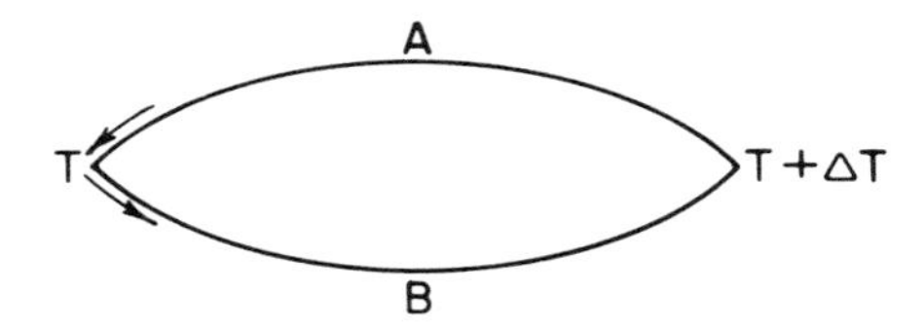

FIGURE 7-1. The Seebeck effect.

generated by various pairs of thermoelements in terms of the temperature of the measuring junction. This is the general approach used in thermoelectric thermometry.

Thermoelectric devices also have been used for the generation of electric current in remote situations. One such application is the generation of power for satellites, or space probes, where battery life would be too short.

It should be noted that most thermoelectric relationships are *not* linear functions of temperature. The slopes of such curves, the changes in emf per degree change in temperature, also called the thermoelectric powers, are not constants. The thermoelectric power provides a useful measure for the description of the response of a given thermocouple over various ranges of temperatures. It is also a convenient means for comparing the thermoelectric properties of different types of thermocouples.

Two thermal phenomena, the Peltier and Thomson effects, will be shown to give rise to the Seebeck effect.

7.2. PELTIER EFFECT

Peltier (1834) showed that when an electric current flows across a junction of two metals (Sb and Bi), heat is liberated or absorbed. He was able to freeze a droplet of water in this way, demonstrating thermoelectric refrigeration. When the electric current flows in the same direction as the Seebeck current, heat is absorbed at the hotter junction and liberated at the colder junction. The Peltier effect is defined as the change in heat content when one coulomb of electricity crosses the junction of two dissimilar materials (Figure 7-2). Metallic materials are used for thermocouples. Semiconductors are employed in thermoelectric refrigeration.

The direction in which the current flows determines whether heat is liberated or absorbed. This effect is reversible and is independent of the shape or dimensions of the conductors composing the junction. It is a function of the compositions of the conductors and the temperature of the junction.

The Peltier effect is quite different from that observed in Joule heating. The latter varies as the square of the current and as the electrical resistance of the conductor. Joule heating is a function of the dimensions of the conductor and does not require a junction of dissimilar materials. In contrast to the Peltier effect, the Joule heat does not change its sign when the current is reversed. It is an irreversible phenomenon. The Peltier effect is reversible and is, as will be shown, quite different from the Joule effect.

The Peltier effect is in no way associated with the nature of the contact between the two conductors. It is not related to "contact potential."

7.3. THOMSON EFFECT

In attempting to describe the behavior of thermocouples in terms of the temperatures and Peltier effects at the two junctions, Thomson (Lord Kelvin) derived a relationship for the emf of a thermocouple based only upon the entropy of the junctions. This predicted that the emf of the thermocouple should be a linear function of the temper-

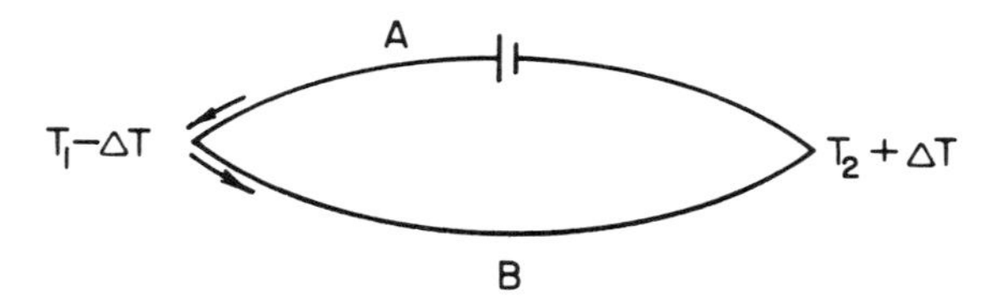

FIGURE 7-2. The Peltier effect.

ature difference between the two junctions. Such a linear response is contrary to the observed nonlinear temperature-emf behavior of all known thermocouples.

Thomson then considered that other reversible thermal changes must occur in thermoelectric circuits in addition to the Peltier effects. These were thought to result from the flow of electric current within the individual thermoelements when a temperature gradient exists along them. Thomson then was able to show (1854) that heat is liberated or absorbed in a single conductor when an electric current flows in the same or opposite direction to the flow of heat. This reversible change in the heat content in a single conductor in a temperature gradient when current passes through it is the Thomson heat.

The Thomson effect is defined as the change in the heat content of a *single* homogeneous conductor of unit cross-section when a unit quantity of electricity flows along it through a temperature gradient of 1 K. Thus, the net heat flow in a conductor per unit volume per unit time is

$$Q = j^2\rho - \sigma j \frac{dT}{dx} \tag{7-1}$$

where j is the current density, ϱ is the electrical resistivity of the conductor, σ is the Thomson coefficient, and dT/dx is the temperature gradient. The first term is the Joule heating and is irreversible. The second term gives the reversible Thomson effect. Thomson noted that σ is the "specific heat of electricity" because it gives the energy change in the conductor per unit current and per unit temperature gradient along the conductor.

The Thomson effect is illustrated in Figure 7-3. The single homogeneous conductor is heated at some point to temperature T_2. A thermal gradient will exist on either side of the heated point. Two points, P_1 and P_2, of equal temperature, T_1, will be found on either side of T_2. If a circuit now is made which includes the single conductor, the temperatures at P_1 and P_2 will be different when the current flows. These temperature changes are the result of the motions of the electrons with respect to the directions of the temperature gradients. The electrons absorb energy in moving against the increasing thermal gradient and increase their potential energy. Heat is absorbed at P_1 where the electron flow direction is opposite to the flow of heat; this causes a decrease in the temperature of P_1. Heat is liberated at P_2, where the direction of the flow of the electrons is the same as the heat flow, and the temperature is increased. These changes in the heat content of the conductor are the Thomson effects. In the case described here, the changes in temperature at P_1 and P_2 are equal and opposite in sign.

7.4. THERMODYNAMICS OF THE THERMOELECTRICITY

Neglecting small convection losses and Joule heating, a thermoelectric circuit can be considered to be a reversible heat engine. The current in a typical thermoelectric circuit is of the order of 10^{-3} A. The electrical resistance of such circuits is usually considerably less than 10 ohms, to achieve maximum sensitivity. The irreversible heat

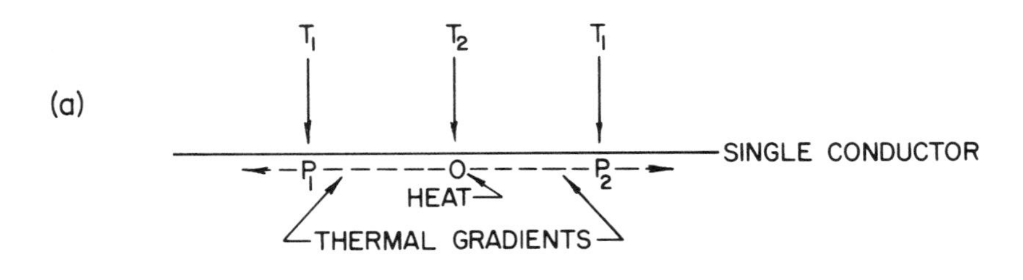

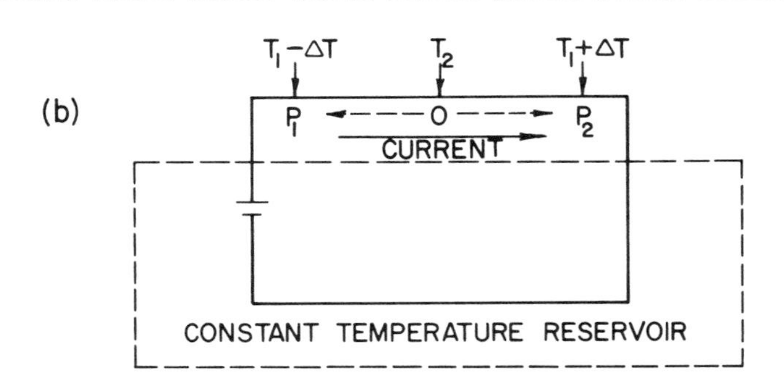

FIGURE 7-3. The Thomson effect.

loss is of the order of 10^{-5} W, or less, which may be neglected. This permits their treatment as being thermodynamically reversible.

Consider the thermoelectric circuit given in Figure 7-1, where the temperatures of both junctions are maintained by individual heat reservoirs. The emf generated in this circuit is E_{AB}. The thermoelectric power is dE_{AB}/dT. Then, the electrical energy is given by

$$E_{AB} = \frac{dE_{AB}}{dt} \Delta T \tag{7-2}$$

It was noted previously that the Peltier effect described the changes in the heat contents of the junctions and that the Thomson effect described the changes in the heat contents in the individual conductors. These thermal changes are:

Peltier Effects (7-3a)
Heat absorbed at the hotter junction $= P_{AB}(T + \Delta T)$
Heat liberated at the colder junction $= -P_{AB}(T)$
Thomson Effects (7-3b)
Heat absorbed in conductor B $= \sigma_B (\Delta T)$
Heat liberated in conductor A $= -\sigma_A (\Delta T)$

Since the thermoelectric circuit can be considered to approximate a reversible heat engine, the thermal and electrical energies may be equated:

$$\frac{dE_{AB}}{dT} \Delta T = P_{AB}(T + \Delta T) - P_{AB}(T) + (\sigma_B - \sigma_A)\Delta T \tag{7-4}$$

Dividing both sides of Equation 7-4 by ΔT gives

$$\frac{dE_{AB}}{dT} = \frac{P_{AB}(T + \Delta T) - P_{AB}(T)}{\Delta T} + \sigma_B - \sigma_A \tag{7-5}$$

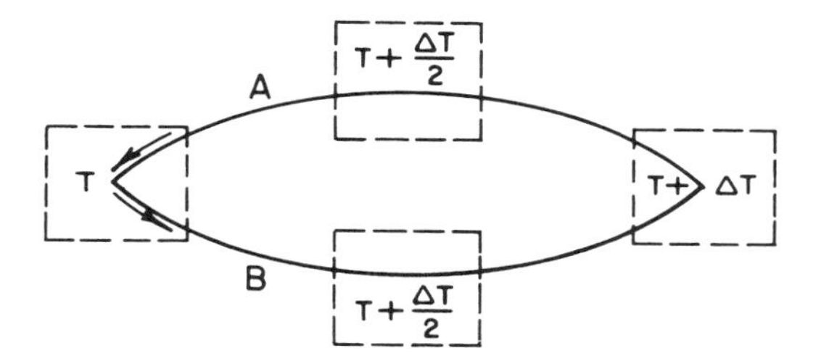

FIGURE 7-4. Same thermoelectric circuit as in Figure 7-1, but with reservoirs added at the midpoints of the conductors.

The fraction on the right is the only term containing the quantity ΔT. It will be recognized as being in the form of a difference quotient, which for the limit where ΔT approaches zero, gives the instantaneous rate of change of the Peltier effect with respect to temperature. When this is substituted into Equation 7-5, the fundamental theorem of thermoelectricity is obtained:

$$\frac{dE_{AB}}{dT} = \frac{dP_{AB}}{dT} + \sigma_B - \sigma_A \tag{7-6}$$

This equation shows that the electrical Seebeck effect is the algebraic sum of the thermal Peltier and Thomson effects.

Now place two additional thermal reservoirs at the midpoints of conductors A and B. Both of these reservoirs are maintained at a temperature which is the mean of those at the hotter and colder junctions (Figure 7-4).

Allow a unit quantity of electricity to flow through the circuit. The net change in the entropy, ΔS, of the reservoirs (surroundings) will be zero for a reversible process. That is

$$\Delta S = \frac{-P_{AB}(T + \Delta T)}{T + \Delta T} + \frac{P_{AB}(T)}{T} - \frac{\sigma_B(\Delta T)}{T + \frac{\Delta T}{2}} + \frac{\sigma_A(\Delta T)}{T + \frac{\Delta T}{2}} = 0 \tag{7-7}$$

When the first two terms on the right are multiplied by $\Delta T/\Delta T$,

$$\Delta S = \left[\frac{-\frac{P_{AB}(T + \Delta T)}{T + \Delta T} + \frac{P_{AB}(T)}{\Delta T}}{\Delta T}\right]\Delta T - \frac{\sigma_B(\Delta T)}{T + \frac{\Delta T}{2}} + \frac{\sigma_A(\Delta T)}{T + \frac{\Delta T}{2}} = 0 \tag{7-8}$$

In the limit, as ΔT approaches zero, the quantity within the brackets becomes

$$-\frac{d}{dT}\left[\frac{P_{AB}}{T}\right]$$

Then,

$$\Delta S = -\frac{d}{dT}\left[\frac{P_{AB}}{T}\right]\Delta T - \frac{\sigma_B(\Delta T)}{T + \frac{\Delta T}{2}} + \frac{\sigma_A(\Delta T)}{T + \frac{\Delta T}{2}} = 0 \tag{7-9}$$

The Thomson effect is defined as the change in heat content for a gradient of $\Delta T = 1$ K. Since T is much greater than 1 K, $T + \frac{1}{2} \approx T$. By means of this approximation

Equation 7-9 becomes

$$\frac{d}{dT}\left[\frac{P_{AB}}{T}\right] = \frac{\sigma_A}{T} - \frac{\sigma_B}{T} \tag{7-10}$$

Performing the indicated differentiation results in

$$\frac{T\dfrac{dP_{AB}}{dT} - P_{AB}}{T^2} = \frac{\sigma_A}{T} - \frac{\sigma_B}{T} \tag{7-11}$$

This becomes, upon simplification,

$$\frac{P_{AB}}{T} = \frac{dP_{AB}}{dT} - \sigma_A + \sigma_B \tag{7-12}$$

Equation 7-6 is used to simplify Equation 7-12 which becomes

$$\frac{P_{AB}}{T} = \frac{dE_{AB}}{dT} \tag{7-13}$$

Thus, the thermoelectric power of a thermocouple is a direct measure of the change in entropy of a thermoelectric junction, since P_{AB} is the change in the heat content of the junction. Equation 7-13 is an important relationship in thermoelectric refrigeration.

Equation 7-13 may be rewritten as

$$P_{AB} = \frac{dE_{AB}}{dT} T \tag{7-14}$$

Upon differentiation with respect to temperature, Equation 7-14 becomes

$$\frac{dP_{AB}}{dT} = \frac{dE_{AB}}{dT} + T\frac{d^2E_{AB}}{dT^2}$$

or

$$\frac{dP_{AB}}{dT} - \frac{dE_{AB}}{dT} = T\frac{d^2E_{AB}}{dT^2} \tag{7-15}$$

Use again is made of the fundamental theorem (Equation 7-6) in Equation 7-15 to give

$$T\frac{d^2E_{AB}}{dT^2} = \sigma_A - \sigma_B$$

or

$$\frac{d^2E_{AB}}{dT^2} = \frac{\sigma_A - \sigma_B}{T} \tag{7-16}$$

Upon integration this becomes

$$\frac{dE_{AB}}{dT} = \int_0^T \frac{\sigma_A - \sigma_B}{T}\,dT = \int_0^T \frac{\sigma_A}{T}\,dT - \int_0^T \frac{\sigma_B}{T}\,dT \tag{7-18}$$

or

$$\frac{dE_{AB}}{dT} = S_A - S_B \qquad (7\text{-}18a)$$

where S_A and S_B are the absolute thermoelectric powers (ATPs) of each of the legs of the thermocouple.

Equation 7-18 can be integrated because the quantity σ/T is entropy, which, from the third law of thermodynamics, approaches zero as T approaches zero. The separation of the thermoelectric power of a thermocouple into the contributions of each of its component conductors provides the fundamental basis for studying any single thermoelement, or any combination of thermoelements, without reference to any other thermoelement. This approach also provides further insight into solid-state phenomena.

7.5. THE CONCEPT OF ABSOLUTE EMF

The separation of Equation 7-18 into two integrals gives rise to the concept known as absolute thermoelectric power, or ATP. The utility of this concept will be apparent.

The concept of ATP, the thermoelectric power of a single conductor, may be difficult to visualize. This may result from the way in which thermoelectric properties most frequently are considered. It is natural to think in terms of the Seebeck effect in which two conductors are required.

One way to overcome this is to consider the Thomson effect. The reversible thermal changes in a single conductor are equivalent to the presence of currents and emfs in the conductor subjected to a thermal gradient. If σ is the "specific heat of electricity", then the heat changes in the conductor are the energy equivalents of the electrical outputs. Thus "absolute" thermoelectric properties are to be expected.

Another way to regard this is to consider a thermocouple made of a superconductor and a metallic conductor. It has been demonstrated that superconducting materials show no electron entropy changes and, therefore, no thermoelectric effects below their superconducting transition temperatures (Section 6.8, Chapter 6). The thermoelectric properties of such a thermocouple, measured below the superconducting transition temperature, can thus be caused only by the normal metallic conductor. The output of this thermocouple must be the absolute emf of the normal metallic conductor. This approach is limited to very low temperatures because the superconducting transition temperatures are all below 24 K, at present.

A third way is to use the element lead (Pb) as a reference thermoelement. This means that Pb serves as the standard, or reference, thermoelement, or leg, of a thermocouple against which the thermoelectric properties of other conductors are established. The ATP of Pb is relatively small compared to that of many other metallic materials. Therefore, the emf of such a thermocouple would be caused almost entirely by the ATP of the other conductor, rather than that of Pb. Values for the ATP of Pb are given in Table 7-1. The use of Pb is limited by its low melting point, 327.4°C.

Platinum has been used widely for reference purposes (see Sections 6.1 and 6.4, in Chapter 6). It can be seen from Equation 7-18 that if the ATP of one thermoelement, such as Pt, is known, and the thermoelectric power of the thermocouple is determined experimentally, the ATP of the unknown leg may be calculated. The ATP of Pt has been well established (Table 7-2).

7.6. LAWS OF THERMOELECTRIC CIRCUITS

The laws of thermoelectric circuits follow directly from Equation 7-18. These are

Table 7-1
ABSOLUTE THERMOELECTRIC POWERS OF LEAD

Temperature (K)	S_{Pb} (μV/°C)
10	−0.43
50	−0.77
100	−0.86
153.2	−1.02
193.2	−1.10
253.2	−1.21
273.2	−1.15
293.2	−1.27

From Pearson, W. B., the effects of chemical impurities and physical imperfections on the thermoelectricity of metals, in *Ultra-High-Purity Metals,* American Society of Metals, 1962, 237. With permission.

Table 7-2
ABSOLUTE THERMOELECTRIC POWER OF SOME ELEMENTS

Temp (K)	Cu	Ag	Au	Pt	Pd	W	Mo
100	1.19	0.73	0.82	4.29	2.00	—	—
200	1.29	0.85	1.34	−1.27	−4.85	—	—
273	1.70	1.38	1.79	−4.45	−9.00	0.13	4.71
300	1.83	1.51	1.94	−5.28	−9.99	1.07	5.57
400	2.34	2.08	2.46	−7.83	−13.00	4.44	8.52
500	2.83	2.82	2.86	−9.89	−16.03	7.53	11.12
600	3.33	3.72	3.18	−11.66	−19.06	10.29	13.27
700	3.83	4.72	3.43	−13.31	−22.09	12.66	14.94
800	4.34	5.77	3.63	−14.88	−25.12	14.65	16.13
900	4.85	6.85	3.77	−16.39	−28.15	16.28	16.86
1000	5.36	7.95	3.85	−17.86	−31.18	17.57	17.16
1100	5.88	9.06	3.88	−19.29	−34.21	18.53	17.08
1200	6.40	10.15	3.86	−20.69	−37.24	19.18	16.65
1300	6.91	—	3.78	−22.06	−40.27	19.53	15.92
1400	—	—	—	−23.41	−43.30	19.60	14.94
1600	—	—	—	−26.06	−49.36	18.97	12.42
1800	—	—	—	−28.66	−55.42	17.41	9.52
2000	—	—	—	−31.23	−61.48	15.05	6.67
2200	—	—	—	—	—	12.01	4.30
2400	—	—	—	—	—	8.39	2.87

From Cusack, N. and Kendall, P., *Proc. Phys. Soc.*, 72, 898, 1958. With permission.

important in many thermoelectric applications. They are significant because they apply to the practical applications of thermocouples in real situations. These include the effects of extension wires between the thermocouple and the instrument, the insertion of other metals or alloys in the thermoelectric circuit, and provide convenient means for the most accurate descriptions of emf-temperature relationships. They also are taken into account in the design of electrical instrumentation.

7.6.1. The Law of Homogeneous Conductors

If two conductors of the same *homogeneous* material are used to form a thermoelectric circuit, the resultant emf will be zero. This results from the fact that σ_A and σ_B (Equation 7-18) are identical. The Law of Homogeneous Conductors states that a thermoelectric current cannot be maintained solely by the application of heat to a single homogeneous conductor regardless of any cross-sectional variations.

Properly selected extension wires from the reference junction of a thermocouple to the measuring device will introduce no extraneous emfs provided that they are exposed to the *same* temperature regime.

If, however, a nonhomogeneous material is employed, the values of σ_A and σ_B will be different. Conductors of this kind will show an emf when exposed to the same temperature regime; thus they introduce extraneous emfs into the circuit and cause erroneous readings. Annealed copper extension wires are commonly used to avoid errors of this type.

Virtually all present-day thermoelectric instruments contain built-in, electronic, reference-junction, temperature-compensating devices. This is done because, in most cases, the reference junctions are at about 20 to 25°C. The standard, practical, reference temperature is 0°C. Therefore, the emfs produced by the thermocouples are for temperature differences which are 20 to 25°C smaller than they would be if the reference junctions had been at 0°C. The automatic, reference-junction, compensating devices add the emf equivalents of about 20 to 25°C to the outputs of the thermocouples so that the instruments give the correct temperature readings.

In other cases, where the reference junctions are at temperatures in the neighborhood of 50°C, miniature, battery-operated, automatic, Peltier-cooling devices may be attached directly to the reference junctions of the thermocouples. These cool the reference junctions to 0°C, with a maximum error of about ± 1°C. The copper extension wires connecting such an assembly to the read-out devices, which are at room temperature, introduce no additional errors because their Thomson coefficients cancel each other.

7.6.2. The Law of Intermediate Conductors

Another consequence of Equation 7-18 is that when no temperature difference exists between the *ends* of a homogeneous conductor, even though temperature gradients may exist between its ends, the emf between the ends will be zero. The application of heat at any point along the conductor will result in temperature gradients of opposite sign. The Thomson effects are not a function of geometry, but only of the temperature difference. The two opposing temperature gradients in the conductor will result in two equal and opposite values for S_A, if the ends are at the same temperature. Thus, any extraneous emf will be nullified.

This behavior is summarized as the Law of Intermediate Conductors which states that the sum of the ATPs of dissimilar conductors is zero when no temperature differences exist between their ends.

A series of many intermediate materials can be made to constitute the measuring junction of a thermocouple. No extraneous emf will be produced if no temperature differences exist between the ends of the materials composing the series.

The same is true for temperature measuring and control devices. These make use of many different metals and alloys in their circuits. These components approach approximately constant temperature conditions after relatively short operating times. Reasonably well-designed instruments will not introduce significant errors because of this.

Some instruments, intended for very accurate measurements of small voltages, are made to guard against small temperature differences. Their measuring circuits are placed in copper-lined, thermostatically controlled containers which are maintained at about 30°C. This ensures that the critical parts of the circuits will be at a constant temperature; thus, normal changes in the ambient temperature will introduce no adverse, extraneous emf effects.

7.6.3. The Law of Successive Temperatures

A third result of Equation 7-18 has been termed the Law of Successive Temperatures. Based upon Equation 7-18a, this may be expressed as:

$$E_{AB} = \int_{T_0}^{T_1} (S_A - S_B)dT + \int_{T_1}^{T_2} (S_A - S_B)dT + \int_{T_2}^{T_3} (S_A - S_B)dT = \int_{T_0}^{T_3} (S_A - S_B)dT \quad (7\text{-}19)$$

In other words, the emf of a thermocouple composed of homogeneous conductors can be measured or expressed as the sum of its properties over successive intervals of temperature.

This concept is employed in the preparation of tables, curves, or equations for emf which best fit various temperature ranges in which a given thermocouple is to be used. Thus, it becomes possible to provide instrumentation for use in specific ranges of temperature. The "memories" of such devices are provided with very accurate emf-temperature data for the desired range (see Equation 7-20).

7.7. APPLICATION OF ATP TO REAL THERMOELEMENTS

The use of ATP enables the examination and understanding of the thermoelectric properties of single thermoelements and pairs forming thermocouples without reference to other thermoelements, such as platinum.

A general type of thermocouple calibration equation given in the literature is of the form:

$$E_{AB} = E_0 + a(T - T_0) + b(T^2 - T_0^2) \quad (7\text{-}20)$$

Here, E_0, a, and b are constants and T_0 is the reference temperature. Such equations may be given for specific ranges of temperature. When this is the case, E_0 represents the emf generated by the thermocouple when its reference junction is at the reference temperature, 0°C, and its measuring junction is at the lowest temperature for which the equation is valid.

This may be explained further by a comparison with Equation 7-19. The term E_0 of Equation 7-20 could be considered to be the equivalent of the first two integrals of Equation 7-19. Then, Equation 7-20 is the equivalent of the third integral and has a range of applicability from T_2 to T_3.

Equation 7-20 can be reexpressed as

$$E_{AB} = \int_{T_0}^{T} (a + 2bT)dT \quad (7\text{-}21)$$

where T_0 again is the reference temperature. It can also be written, using Equation 7-18a, as

$$\frac{dE_{AB}}{dT} = a + 2bT = S_A - S_B \tag{7-22}$$

If the ATPs of the thermoelements A and B are given by the linear functions

$$S_A = \alpha + m_A T \tag{7-23a}$$

and

$$S_B = \beta + m_B T \tag{7-23b}$$

then,

$$\frac{dE_{AB}}{dT} = S_A - S_B = (\alpha - \beta) + (m_A - m_B) T \tag{7-24}$$

By comparing Equation 7-24 with Equation 7-22, it is seen that

$$a = \alpha - \beta \tag{7-25a}$$

and

$$2b = m_A - m_B \tag{7-25b}$$

Thus, the constants of the original equation have special meanings in terms of the ATPs of the components of the thermocouple. In other words, the coefficients in an equation of the type given by Equation 7-20 are more than just curve-fitting parameters. The slopes, m_A and m_B, are explained in elementary quantum mechanic terms in the following sections.

For the case of many thermocouples in common use, which have nearly linear temperature-emf properties, the coefficient b must be very small. This means that the slopes m_A and m_B of S_A and S_B must be very nearly equal. Very few combinations of thermoelements show this desirable behavior. This is one of the reasons why so few combinations of thermoelements are in common use.

Equation 7-20 has a term containing $(T^2 - T_0^2)$. This means that E_{AB} is a parabolic function of the temperature difference. The smaller this term is, the smaller will be the departture of E_{AB} vs. T from linearity. The more linear the emf is with temperature, the more desirable the thermocouple is considered to be by instrument manufacturers. Components, such as slidewires, and electrical circuits, based upon nonlinear responses of sensing devices are expensive to produce because they are not as readily made as linear ones.

The ideal thermocouple would consist of thermoelements whose ATPs were exactly parallel functions of temperature. Here $m_A = m_B$. Then

$$S_A - S_B = \alpha - \beta = a$$

and

$$E_{AB} = \int_{T_0}^{T} a dT = a(T - T_0) \tag{7-26}$$

which is a linear function of the temperature difference. It has not been possible to produce such a thermocouple (Section 7.3).

From a geometric point of view, the general behavior of a pair of thermoelements may be pictured as given in Figure 7-5. The thermoelectric power of a thermocouple composed of thermoelements A and B may be read directly from the figure. According to Equation 7-18a, the thermoelectric power at any temperature is just the algebraic difference between S_A and S_B, as shown in the figure for the two temperatures T_1 and T_2. As drawn, the figure corresponds to Equation 7-20. If the curves for S_A and S_B were parallel, the figure would correspond to Equation 7-26. In other words, the emf generated by a pair of thermoelements, in the form of a thermocouple, is the area between the two curves over the temperature range between the reference temperature and the temperature being measured.

7.8. MODELS FOR THERMOELECTRIC BEHAVIOR

The foregoing presentations, based upon thermodynamic properties, describe the observed behaviors of thermoelectric phenomena. They provide no model which explains the electron mechanisms involved. Greater insight may be obtained by another approach which sheds light upon the solid-state physical reactions which determine thermoelectric properties.

In discussing the Thomson effect (Section 7.3), note was made of the absorption and release of energy by the electrons as they flowed along the conductor either against or with the direction of heat flow. It was also shown (Equation 7-1) why σ has been described as the "specific heat of electricity". This idea is the basis for one of the following approaches.

The following sections, in essence, are devoted to derivations of the slopes, m_i of Equation 7-25b, the factors that affect them and the understandings they provide. Such concepts afford additional insight into electron behavior in solid-state phenomena. In addition, they help to explain the properties of the commercially available thermoelements and why so few are in common use.

7.8.1. Normal Metals

Consider a single normal metal conductor in a known, constant temperature gradient with no electric current flowing. The change in the Gibbs free energy is given by

$$\Delta F = V\Delta P - S\Delta T \tag{7-27}$$

in which V is the volume, P is the pressure, and S is the entropy. At constant pressure, $\Delta P = 0$, so Equation 7-27 becomes

$$\Delta F_1 = -S_1\Delta T \tag{7-28}$$

for the case where no current flows through the conductor. Or, expressing the entropy in terms of heat capacity,

$$\frac{\Delta F_1}{\Delta T} = -\int_{T_1}^{T_2} \frac{C_P}{T}\,dT \tag{7-29}$$

where T_1 and T_2 define the temperature gradient.

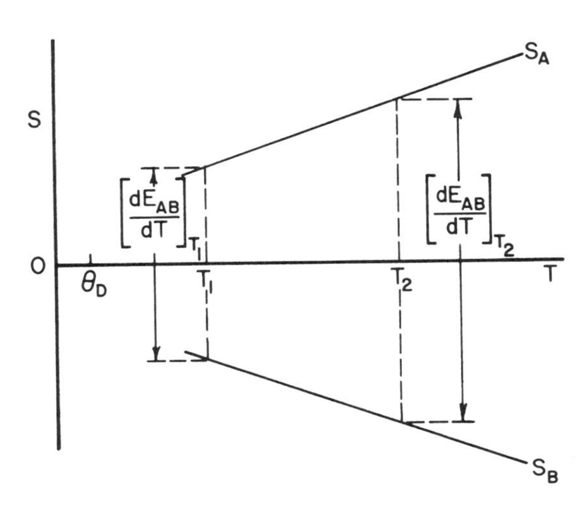

FIGURE 7-5. Schematic diagram of the ATP of two thermoelements.

If electricity now is allowed to flow in the single conductor maintained in the same temperature gradient, its heat content, and consequently its free energy, will change in accordance with the Thomson effect. Accordingly, the free energy of the conductor under these conditions will have a gradient expressed by

$$\frac{\Delta F_2}{\Delta T} = -\int_{T_1}^{T_2} \frac{C_P'}{T}\, dT \tag{7-30}$$

The difference between the gradients in free energy can be defined as

$$\frac{dF}{dT} = \frac{\Delta F_2}{\Delta T} - \frac{\Delta F_1}{\Delta T} = \int_{T_1}^{T_2} \frac{\Delta C_P}{T}\, dT \tag{7-31}$$

Now, based on the approximation given by Equation 4-7,

$$\Delta C_P = \Delta C_V \tag{7-32}$$

Equation 7-31 becomes, upon substitution,

$$\frac{dF}{dT} = \int_{T_1}^{T_2} \frac{\Delta C_V}{T}\, dT \tag{7-33}$$

Assuming essentially free electron behavior, the heat capacity of a normal metal can be approximated above θ_D from Equations 5-52a and 5-53b as

$$C_{V1} = 3Nk_B + A\frac{\pi^2}{2}\frac{nNk_B{}^2T}{E_F} \tag{7-34}$$

where A is a conduction coefficient which accounts for variations in the electron conduction model between various metals and the effects of alloying upon metals (see

Sections 5.6.1 and 5.6.2 and Table 5.5 in Chapter 5, Volume 1 and 7.7.6,) n is the valence and N is Avogadro's number.

In the case where current is flowing, the motion of the electrons near E_F is no longer random, and in the limit, C_{V2} can be approximated by $3Nk_B$. The difference between the two cases is given by

$$\Delta C_V = -A\frac{\pi^2}{2}\frac{nNk_B^2T}{E_F} \tag{7-35}$$

When this is substituted into Equation 7-33, and the expression integrated between the limits of $T_1 = 0$ and $T_2 = T$, it is found, to a first approximation, that

$$\frac{dF}{dT} = -A\frac{\pi^2}{2}\frac{nNk_B^2T}{2E_F} \tag{7-36}$$

If, alternatively, it is considered that the change in the gradient in free energy is due only to the flow of electrons, and it is assumed that such current flow is a reversible process, by analogy to the thermodynamics of reversible cells,

$$\Delta F = -n\mathfrak{F}\Delta E \tag{7-37}$$

in which F is Faraday's number and ΔE is the emf across the cell. Since F = Ne, Equation 7-37 can be written as

$$\Delta F = -nNe\Delta E$$

and, when both sides are divided by ΔT, this becomes

$$\frac{\Delta F}{\Delta T} = -nNe\frac{\Delta E}{\Delta T} \tag{7-38a}$$

For very small intervals, this may be expressed as

$$\frac{dF}{dT} = -Nne\frac{dE}{dT} \tag{7-38b}$$

Equations 7-36 and 7-38b may be equated to give

$$-Nne\frac{dE}{dT} = -A\frac{\pi^2}{2}\frac{nNk_B^2T}{E_F}$$

or, when simplified,

$$S = \frac{dE}{dT} = A\frac{\pi^2}{2}\frac{k_B^2T}{eE_F} \tag{7-39}$$

This gives the ATP of a conductor in terms of the Fermi level and the temperature; the other factors are constants for a given material. Under a given set of conditions, the only variable is E_F.

Equation 7-39 is valid only for $T \geqslant \theta_D$ since it is based upon Equation 7-34. It is applicable for the cases where nearly free electrons are the carriers, such as the normal metals.

7.8.2. The Mott and Jones Model for Normal Metals

The Mott and Jones model is based upon a purely quantum mechanic approach and represents an important contribution. Since the detailed derivation by these authors is beyond the scope of the present text, it only will be outlined briefly here. It will be apparent that the fundamental relationship from which the ATP is derived is different from that used to arrive at Equation 7-39.

They considered a rod of unit cross-section in which there is an electric current j, a heat current Q, a temperature gradient $\partial T/\partial x$, and an electric field F. The energy developed per unit volume, per unit time is

$$U = Fj - \frac{\partial Q}{\partial x} \tag{7-40}$$

The quantities j and Q are evaluated by similar integrals. The integral for j sums the contributions of all electrons over all wave-vector space in a form equivalent to nev_x. The integral for Q sums the heat energy in a form equivalent to nEv_x, where n is the number of electrons, e is the charge on an electron, v_x is the velocity of an electron in the direction of the temperature gradient, and E is its energy. When the quantity F is eliminated, and the irreversible heat is taken into account, it is found that a first approximation of the general expression for the ATP of a metal is given by

$$S = \frac{\pi^2 k_B^2 T}{e} \left[\frac{\partial \ln \phi(E)}{\partial E}\right]_{E = E_F} \tag{7-41}$$

Here, $\phi(E)$ is given by

$$\phi(E) \simeq \text{constant}\ \tau(\bar{k}_x)\, N(E) \tag{7-42}$$

in which $\tau(\bar{k}_x)$ is the relaxation time as a function of the wave vector, k_x, and N(E) is the density of states. The form of Equation 7-42 recalls that of Equation 5-64. This was given as

$$\sigma(E_F) = \frac{n(E_F)e^2}{m} \tau(E_F) \tag{5-64}$$

If e^2/m is considered to be a constant in Equation 5-64, then

$$\sigma(E_F) = \text{constant}\ n(E_F)\tau(E_F) \tag{7-43}$$

When provision is made for the dependence of τ in Equation 7-42, it is shown that

$$\phi(E)_{E=E_F} = \sigma(E_F) \tag{7-44}$$

Thus, the Mott and Jones model is based upon electrical conductivity. An approximation is made in the range $E \simeq E_F$, where the parameter x includes the effects of $\bar{k}_x$ and $\tau(\bar{k}_x)$, such that

$$\sigma(E) = \text{constant}\ E^x \tag{7-45}$$

When this is substituted into Equation 7-41, it is found that

$$S = \frac{\pi^2 k_B^2 T}{3eE_F} x = -\frac{2.45 \times 10^{-2} T}{E_F} x \ (\mu V/°C) \tag{7-46}$$

This relationship is the same as that given by Equation 7-39; here, the electron parameter is $\pi^2 x/3$ instead of A. It, too, is valid only above θ_D (see Section 7.8.6).

For the noble metals, x is found by experiment to be about −3/2 (see Section 7.8.6). The ATP for the noble elements is given by

$$S = -\frac{\pi^2}{2} \cdot \frac{k^2{}_B T}{eE_F} \ (\mu V/°C) \tag{7-47}$$

When the charge on the electron is taken into account, Equation 7-47 becomes positive.

Equation 7-46 also is valid for other groups of metals, provided that suitable values are used for the parameter x. The transition elements are treated separately (see Sections 5.6.1 and 5.6.2 and Table 5-5 in Chapter 5, Volume I, and Section 7.8.4).

7.8.3. Transition Elements

Equation 7-39 is valid only for normal metals because of the assumptions implicit in Equation 7-34. In normal metals, the electrical conduction is by the electrons. In the case of transition elements, such as iron, cobalt, and nickel, there is a positive hole contribution both to the electrical conductivity and to the heat capacity. This arises from the fact that such elements have incompletely filled d levels (Section 5.6.2 in Chapter 5, Volume I). Here, where $(E_o - E_F)$ represents the energy range of the incompletely filled portion of the d band (Figure 5-9, Chapter 5, Volume I) the heat capacity is given, based upon a derivation similar to that for Equation 5-55, and it is found that

$$C_V = 3Nk_B + A' \frac{\pi^2 nNk_B{}^2 T}{6(E_o - E_F)} \tag{7-48}$$

When this is treated as in Section 7.8.1, the ATP of pure transition elements is found to be

$$S = -A' \frac{\pi^2 k_B{}^2 T}{6e(E_o - E_F)} \tag{7-49}$$

It should be noted that Equation 7-27, which forms the basis for these derivations, was simplified by considering ΔP to be constant at one atmosphere. This means that Equations 7-39 and 7-49 are valid at a constant pressure of one atmosphere. This includes almost all applications of thermocouples.

When thermoelements are subjected to appreciable pressures, this factor cannot be neglected. Many thermoelements show an appreciable sensitivity to this variable. For example, a pressure of 50 kbar can induce an error of about ±2% in thermocouple readings in the range from 1200 to 2000°C; the sign of the error depends upon the thermoelements involved. The effects of pressure cannot be predicted readily from the Mott and Jones relationships because they are not based directly upon thermodynamic relationships.

In some cases, such as the injection molding of metals and plastics, the temperature control may be sufficiently critical as to require that pressure-induced errors be taken into account. The pressures involved may range from about 500 to 2000 atm.

7.8.4. The Mott and Jones Model for Transition Elements

The behavior of the d-level holes must be taken into account when assessing the behavior of the factor $\ell n\phi(E)$ in Equation 7-41. Since the d bands of the transition elements are not filled with electrons, the behaviors of these elements can be described by considering the behavior of their unfilled states, or holes. On this basis, it is found that

$$N_d(E) \propto (E_o - E_F)^{1/2} \tag{7-50}$$

If it is assumed that the factor $N_d(E)$ is the major variable in Equation 7-43, and if the probability of s – to – d transitions is taken into account in terms of $N_d(E)$, then the ATP of a pure transition thermoelement is found to be

$$S \simeq -\frac{\pi^2}{6} \cdot \frac{k^2{}_B T}{e(E_O - E_F)} \qquad (7\text{-}51)$$

Except for the coefficient, this is the same relationship as that given by Equation 7-49. Since $N_d(E)$ decreases sharply as E_F approaches E_O, it is expected that the ATP should be large and negative compared to Equation 7-47. The sign of e is taken to be positive here, since $(E_O - E_F)$ represents a range of empty states, or holes.

In both Equations 7-49 and 7-51, significant changes in $(E_O - E_F)$ can result from contamination and/or impurities in the thermoelements and affect S (see Sections 7.9.3 and 7.9.4). Therefore, determinations of Fermi levels, made on the basis of these equations, may be inaccurate if these impurities are not taken into account.

7.8.5. Comparison of Models

It is instructive to examine and compare the results of both of the models presented here. The model based upon the contribution of the electrons to heat capacity gave for the normal, or noble, metals

$$S = A\frac{\pi^2 k_B{}^2 T}{2eE_F} \qquad (7\text{-}39)$$

and for transition elements,

$$S \simeq A'\frac{\pi^2 k_B{}^2 T}{6e(E_O - E_F)} \qquad (7\text{-}49)$$

The Mott and Jones theory, based upon electrical conductivity, gave for the noble metals

$$S = \frac{\pi^2}{3} \cdot \frac{k^2{}_B T}{eE_F} \cdot x \qquad (7\text{-}50)$$

and for transition metals

$$S \simeq -\frac{\pi^2}{6} \cdot \frac{k^2{}_B T}{e(E_O - E_F)} \qquad (7\text{-}51)$$

Both sets of results are approximations. It can be seen by comparing Equations 7-39 and 7-50 that

$$\frac{A\pi^2}{2} = \frac{\pi^2}{3}x \quad \text{and} \quad A = 2x/3 \qquad (7\text{-}52)$$

A good approximation for the electron contribution to the heat capacity of a normal metal of valence n is given by

$$C_{ve} = \frac{\pi^2}{2} \cdot \frac{nNk^2{}_B T}{E_F} \qquad (7\text{-}53a)$$

When this is compared with Equation 7-34 it is seen that

$$\frac{A\pi^2}{2} = \frac{\pi^2}{2} \quad \text{and} \quad A = 1 \tag{7-53}$$

If the empirically determined approximate value $x \simeq -3/2$ for noble metals, is used in Equation 7-52

$$A = \frac{2}{3} \cdot -\frac{3}{2} = -1$$

Equation 7-39, then, requires an adjustment by a factor of −1. A comparison of Equations 7-49 and 7-51 also shows that

$$A' \simeq -1 \tag{7-53a}$$

The conduction coefficients A, A′, and x depend upon the way in which the electron contribution to the heat capacity varies, or the way in which the electrical conductivity varies. In other words, these parameters depend upon the variations in the electron models needed to describe the different metallic elements. The differences in the values of these conduction coefficients for different metals and for their alloys constitute an important reason for the inability to secure thermoelements with parallel slopes; these would give linear emf characteristics with temperature (see Equation 7-26 and compare it with Equation 7-20). Additional insight into the coefficients x, A, and A′ is provided in Section 7.8.6 and in conjunction with that given in Sections 5.6.1 and 5.6.2, and in Table 5-5, Chapter 5, Volume I.

7.8.6. Thermoelectric Power and Heat Capacity

It is interesting to compare the relationships for the carrier contributions to the heat capacity of solids with the corresponding equations for the ATP of the solids. For noble metals:

$$C_{Ve} \simeq \frac{\pi^2}{2} \cdot \frac{nNk^2{}_BT}{E_F} \quad \text{and} \quad S \simeq \frac{\pi^2}{2} \cdot \frac{k^2{}_BT}{eE_F}$$

For transition metals

$$C_{Vh} \simeq -\frac{\pi^2}{6} \cdot \frac{nNk^2{}_BT}{(E_o - E_F)} \quad \text{and} \quad S \simeq -\frac{\pi^2}{6} \cdot \frac{k^2{}_BT}{6\,e(E_o - E_F)}$$

It can be seen from the above that S is given by the heat capacity of a single carrier divided by its charge:

$$S = \frac{C_V\,(\text{carrier})}{e\,(\text{carrier})} \tag{7-54}$$

This relationship offers insight into the thermoelectric behavior and other properties of many metals and alloys. It also provides a means for the calculation of the contribution of the carrier to the heat capacity of the solid. This, in addition, permits the determinations of coefficients such as those given in Equations 7-52, 7-53, and 7-53a. These may be used as indexes of the differences in the conduction mechanisms of the carriers in metals and their alloys. The parameter x in Equation 7-46 is different for each element, even for those of a given group. For example, Mott and Jones give values of x equal to −1.4, −1.6, and −1.1 for copper, silver, and gold, respectively. These may be converted into the parameter A by Equation 7-52 for use in Equation

7-39. Variations also occur in the electron conduction coefficients A′ of Equation 7-49. Consequently, even though alloys can be made with virtually identical values of E_F or $(E_O - E_F)$, their individual slopes will not be equal because of the variations in the electron conduction coefficients (see Sections 5.6.1 and 5.6.2 and Table 5-5, in Chapter 5, Volume I).

7.8.7. Comparison of Thermoelectric Powers

An examination of Equations 7-47 and 7-51 shows why almost none of the normal, or noble, metals are used as thermoelements at elevated temperatures. At the same time it gives the reason why nearly all of the thermoelements in common use consist of a certain few transition elements or their alloys.

A comparison of the two classes of metals can be based upon their Fermi energies. The Fermi levels of the noble metals are about 6eV. The transition elements used in thermoelectric applications and their dilute alloys (up to about 20% alloying elements) have values of $(E_O - E_F)$ of approximately leV. On this basis, the ATP of a transition metal element is at least twice that of a noble metal thermoelement. A thermocouple made of transition metals, or alloys, would show a thermoelectric power of at least four times that of a noble-metal thermocouple at a given temperature. In addition, certain pairs of transition thermoelements have relatively small coefficients of squared term of Equation 7-20; their electron conduction coefficients are very similar. Thus, they come close to providing ideal thermoelectric behavior. From the foregoing it should be apparent why so few pairs of thermoelectric materials are in common use, and why nearly all of these are transition elements or their alloys: they show relatively large, nearly linear emf characteristics vs. temperature (see Section 7.7).

7.9. FACTORS AFFECTING THE FERMI LEVEL

The foregoing shows that the Fermi level is the principle factor in the relationship for ATP. An understanding of the behavior of this property is necessary for thorough understanding of thermoelectric phenomena. It also will provide insight for understanding alloy phases (Chapter 10, Volume III).

The theoretical developments assumed that the conductors were ideal, i.e., that they were perfect, continuous, homogeneous, isotropic media with no impurities. Real conductors are far from this ideal. Their lattices are neither perfect nor isotropic. Almost invariably they are polycrystalline, contain appreciable amounts of impurities and/or alloying elements, and are not entirely free from internal stresses. Questions naturally arise regarding the effects of these factors upon the ATPs of the conductors.

The most important factors affecting the Fermi level will be examined individually, primarily to illustrate their effects upon the ATP. It must be emphasized that the changes in E_F also must be taken into consideration in every case in which it is a factor, such as in alloy phase formation.

7.9.1. Temperature

The effect of temperature upon the Fermi level is given by Equation 5-26. In discussing this equation, it was shown that for most purposes

$$E_F(T) \simeq E_F(0) \tag{7-55}$$

This is true primarily for normal elements and their solid-solution alloys. It also holds for dilute alloys of the transition elements containing up to about 20% of alloying elements in solid solution.

However, when the d bands of transition elements approach maximum filling as a

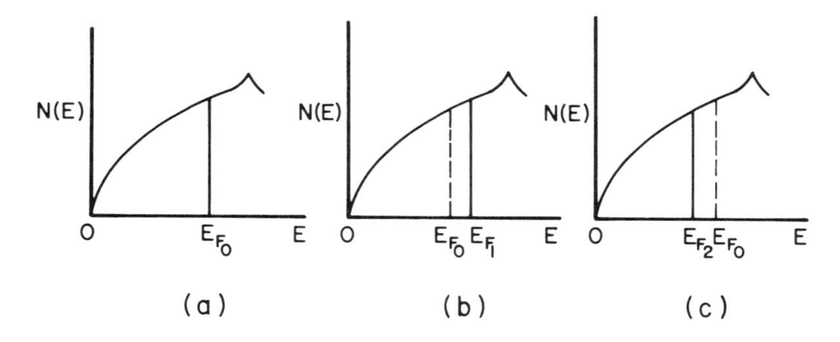

FIGURE 7-6. Schematic representations of the densities of states for dilute alloys of noble metals.

result of alloying, the small changes in E_F cannot be neglected. This will be discussed below in Section 7.9.4. With this exception, the Fermi levels of other alloys can be considered to be virtually invariant with temperature with a high degree of reliability.

7.9.2. Alloying Elements in Solution in Normal Metals

It was shown previously (Equation 5-24) that the Fermi energy can be considered to be a function of the electron:ion ratio. Equation 5-24 may be rewritten as

$$E_F = \text{constant}\left[\frac{N}{V}\right]^{2/3}$$

since, in the present case, all of the other factors are constants for a given metal or alloy. This behavior may be presented graphically (Figure 7-6) based upon the curve of the density of states for a noble metal, or for a trivalent metal.

The curve for the pure noble metal (Figure 7-6a) shows a half-filled zone. This must be true because the noble metals have one electron per atom and the zone can accommodate two electrons per atom. The electron: ion ratio equals unity.

Now consider a dilute, random, substitutional, solid solution in which the solute atom has a greater number of valence electrons than the noble metal of the lattice in which it is dissolved. If the added element has a valence of Z_β, each atom in solution will add $(Z_\beta - 1)$ additional electrons to the lattice. The electron:ion ratio will be greater than unity; the Fermi level of this alloy, E_{F1}, will be expected to be greater than that (E_{F0}) of the pure, noble metal (Figure 7-6b). This causes the ATP of the alloy to be smaller than that of the pure, base metal at a given temperature.

The situation in which the added element contains fewer valence electrons than the atom of the host lattice is shown in Figure 7-6c. If the added element has a valence of Z_β, each such atom in solution will create a deficit of $(Z_\beta - 1)$ electrons. The electron:ion ratio will be less than unity, and the Fermi level would be expected to decrease. Here, the value of S would be larger than that of the pure, base metal at a given temperature.

In both of the above cases the assumption is made that the shape of the zone is unaffected by the presence of dilute amounts of alloying elements. In addition, it is assumed that the Fermi surfaces are approximately spherical and unaffected by the zone wall. Also, in both of these cases, the change in the Fermi level would be proportional to the amount of alloying elements present in solution in the noble, or normal, metal because this determines the electron:ion ratio, and, consequently, affects E_F.

7.9.3. Dilute Solutions in Transition Metals

The assumption again must be made that dilute amounts of alloying elements in substitutional solid solution in transition metals do not appreciably change the shapes of the zones of the pure elements and that the spherical Fermi surfaces are unaffected by the zone boundaries. In the case of transition metal solutes, account must be taken of both s and d states (Section 5.6.2), since the overlap of the bands must be considered to explain their properties. As shown in Figure 5-9, the energy range from E_F to E_0 is the range of d-level holes, or absence of electrons. For example, Fe has 2.2 holes per atom, Co has 1.7, and Ni has 0.6 in the crystalline state.

When normal atoms with completed inner electron levels, such as noble metals, are in solid solution in transition metals, their valence electrons tend to occupy states in the range of d-level holes of the transition element. This diminishes the energy span at the top of the d band; the value of $(E_o - E_F)$ for the alloy becomes smaller than that of the unalloyed transition element and S becomes greater than that of the host element at a given temperature. Transition elements in substitutional solid solution in other transition elements can have the effect of adding holes to those already present. The energy range at the top of the band thus can become greater than that of the pure element. Here, the ATP of the alloy is less than that of the unalloyed base at a given temperature.

In a given dilute alloy, $(E_0 - E_F)$ is essentially constant. This comes about because E_F is virtually constant with temperature and $(E_0 - E_F)$ is reasonably large, that is, much greater than k_BT. It can be seen that small changes in E_F will cause only small changes in $N_d(E)$ because it lies on the flatter portions near the top of the curve. Thus, minor changes, or variations, in $N_d(E)$ in Equation 7-43 will be negligible when included in Equation 7-41. This is valid for alloy concentrations up to about 20%.

7.9.4. Concentrated Solutions in Transition Metals

In the case of concentrated solid solutions of normal or noble metals in transition elements, in such systems as Cu-Ni, Ag-Pd, and Au-Pd, in which the ratio of noble atoms to transition atoms approaches 60/40, the d levels approach maximum filling. This is shown schematically in Figure 7-7.

The condition of maximum filling implies that $(E_0 - E_F)$ is quite small. Therefore, the small variations in E_F as a function of temperature cannot be neglected, as was the case for the dilute alloys; such changes now may constitute an appreciable percentage of $(E_0 - E_F)$. More significantly, it will be recalled (Section 7.8.4) that the principle variable was the density of states of the d band. As indicated in Figure 7-7, a small change in E_F will result in a relatively large change in $N_d(E)$. Such changes obviously will be reflected when Equation 7-43 is used in Equation 7-41. Thus, $(E_0 - E_F)$ cannot be regarded as a constant for alloys in which the d band is almost completely filled.

Another departure from the prior discussions resides in the fact that it is no longer possible to assume that the shapes of s and d bands, or those of the Brillouin zones, are unaffected by the presence of the high concentrations of solute atoms, as was the case for dilute solutions. It is possible for this model only to show the extent to which the d levels appear to be unfilled.

This model is oversimplified since electrons with energies near the top of the d band are hybrid s-d states. The distributions of such electrons between the s and d levels are not known exactly, but are determined by a probable division between them. One approach has been to consider that the electrons of solute and solvent atoms share common s and d bands in alloys. This leads to a band model, not unlike that shown in Figure 7-7, known as the "rigid-band model".

The rigid-band model permits an approximation of the electron behavior of alloys as the d band fills. When E_F approaches E_O in the shared d band, the value of the Fermi function decreases and becomes quite small. This means that the probability of finding any additionally added electrons in the shared d levels decreases and becomes

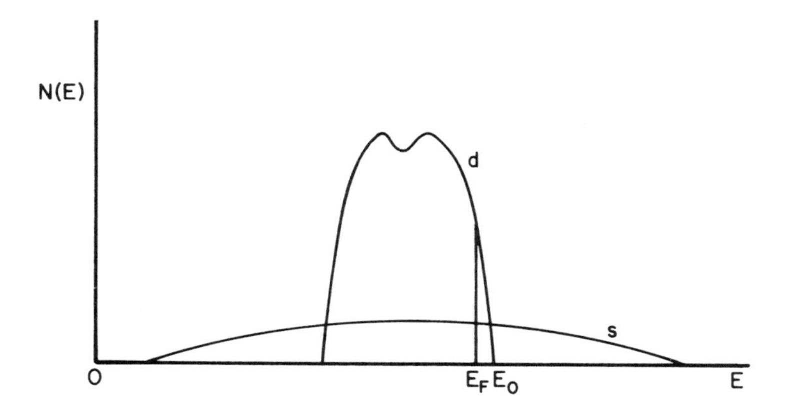

FIGURE 7-7. Schematic representation of the densities of states of the d band, $N_d(E)$, and of the s band, $N_s(E)$, for concentrated solid solutions of transition elements.

small. The probability of finding such electrons in shared s states is large because of the overlap of the s and d bands. Thus, an electron concentration is reached at which the shared d levels are filled to a maximum degree. It appears that such d levels do not become filled completely. At electron concentrations greater than that of the maximum filling of the d levels, the electron behavior becomes increasingly more s-like as more electrons are added. As increasing s-like behavior occurs, the magnitude of ($E_O - E_F$) increases. This is explained by considering that the s band (Figure 7-7) shifts to slightly lower energies when this occurs.

As the alloy content causes the d band to approach maximum filling, ($E_O - E_F$) approaches a minimum value; the ATP increases accordingly. Beyond the composition of such filling, ($E_O - E_F$) increases and the ATP decreases. An isothermal plot of the ATP as a function of composition, therefore, is approximately parabolic in the range of the maximum filling (see Section 7.14.2).

7.9.5. Stress or Working

The effects of stress or mechanical work must be considered in terms of lattice distortion. The resulting disarray "warps" the band structure and the Brillouin zone and increases the scatter of electrons. This also results in changes in the Fermi surface, which may no longer be regarded as being spherical, and is reflected in changes in the ATP.

The types of stress, their magnitudes, and their distributions have been shown to affect the thermoelectric properties. The characters of the residual stresses which usually are encountered after cold working are mixed tensile and compressive combinations, with at least some degree of triaxiality, depending upon the type of working. Under such conditions, changes in E_F are difficult to assess. The separation of the effects of such variables has not yet been achieved. Generally, the lumped effects of the complex stress distributions, resulting from the different mechanical treatments, are considered as representing different "states". Thus far it has not been possible to improve upon such a description.

This becomes increasingly apparent when the deformation textures of metals and alloys are examined. Under such conditions, a moderately cold-drawn wire will produce a small emf relative to an annealed wire of identical composition. This indicates that the Fermi level has changed as a result of the stored energy in the deformed lattice.

It must also be remembered that under optimum annealing conditions, polycrystalline materials will contain imperfections as well as dislocation densities of from 10^6 to 10^8/cm^2. In addition, most so-called "pure" metals actually contain large numbers of impurity ions, even though the percentages of such ions present are quite low. So, even under the best of conditions, the values obtained for E_F or $(E_O - E_F)$ will be influenced by the impurities and imperfections present in all polycrystalline materials.

7.10. ATP OF NOBLE METAL ALLOYS

In the preceding discussions it was assumed that the shape of the band or Brillouin zone is virtually unaffected by the presence of dilute amounts of alloying elements in solid solution. This led to the approximation that E_F is a function of the composition of the solution, the electron:ion ratio being a function of composition, as shown in Section 7.9.2. The electrical conductivity is expected to be sensitive to variations in E_F, (Equation 5-64). Thus, the electrical conductivity as a function of the Fermi level, $\sigma(E_F)$, should be proportional to the electrical conductivity as a function of composition, $\sigma(C)$.

A modification of Equation 7-41 may be used to approximate this behavior:

$$S = \text{const}_1 \frac{\partial}{\partial E} [\ell n \sigma(E)]_{E=E_F} \tag{7-56}$$

Since the relationship of E_F to composition has been shown, and the proportionality between $\sigma(E_F)$ and $\sigma(C)$ is considered, it can be approximated that

$$S = \text{const}_2 \frac{\partial}{\partial C} [\ell n \sigma(C)] \tag{7-57}$$

When the indicated operation is performed, and note is made of the reciprocal relationship between electrical resistivity, $\varrho(C)$, and conductivity,

$$S = \text{const}_3\, \rho(C) \frac{d\sigma(C)}{dC} \tag{7-58}$$

where C is the amount of the alloying element, in atom percent, in solid solution. The variables may be calculated using Equations 6-24, 6-25, and 6-32.

It should be noted that Equation 7-58 gives negative values for the ATP since $d\sigma(C)/dC$ is negative. This is shown in Figure 7-8. Here, the ATP of a pure, noble metal is shown as being small and positive. Small additions of binary components cause the alloys to become increasingly more negative.

This response is in agreement with the data given in Table 7-3. The data show that noble atoms in solution in other noble metals cause only a slight decrease in ATP. Such behavior can be explained on the basis that the electron:ion ratio remains unchanged and that small changes in the coefficients given in Equation 7-52 which result from changes in conduction and in the shape of the zone give rise to the small ATP. Equation 7-58 also explains this in other terms. The change in resistivity is small; the change in conductivity also is small and negative. Thus, S is expected to be small and negative.

Multivalent normal solute elements in noble metals should increase E_F, (Figure 7-6) which, in turn, should decrease S (Equation 7-39). This is in agreement with the behavior noted in the table.

Transition elements in solution in the noble metals should give the greatest changes in the ATP. These elements, with vacant d levels, are strong scattering centers; therefore, they exert large effects upon $\varrho(C)$ and $d\sigma(C)/dC$, giving larger, negative values for the ATP.

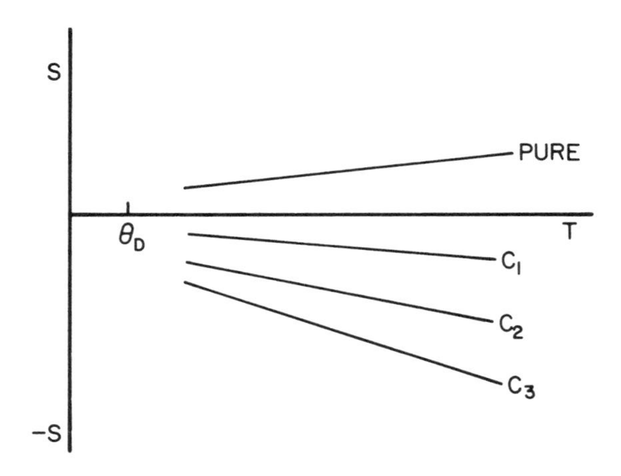

FIGURE 7-8. Schematic diagram of the effects of small, binary alloying additions, C_i, in solid solution in a noble metal ($C_1 < C_2 < C_3$).

In terms of Equation 7-39, transition elements in solid solution in noble metals should cause some of the electrons, which normally would occupy s states, to occupy the d states belonging to the transition atoms. This effectively decreases electron:ion ratio and lowers E_F. A larger (more negative) ATP consequently results.

7.11. ATP ALLOYS OF MULTIVALENT METALS

Alloy solutes in multivalent normal metals, whose inner electron shells are completed, behave similarly to noble-metal solutes. Added alloying ions in solid solution with charges less than that of the solvent lower E_F and increase S. Ions with charges greater than that of the solvent increase E_F and thus decrease S. Both behave in a way similar to that shown in Figure 7-6 b and c.

The presence of transition ions in solid solution in multivalent solvents also results in behavior similar to that noted for them when in noble-metal solvents. These solutes provide d orbitals for the s and p valence electrons and effectively lower E_F. This is evidenced by increases in the values of S, as is shown in Table 7-3.

7.12. ATP OF ALLOYS OF TRANSITION ELEMENTS

Alloys of transition elements are described conveniently by considering two general classifications. One of these concerns alloying additions of normal metals with complete inner electron shells, including the noble metals. The other deals with added alloying transition elements in solid solution in other transition element bases. The explanations given in Sections 7.9.3 and 7.9.4 provide the bases for the following discussions.

7.12.1. Solute Ions with Completed Electron Shells

In this case the valence electrons from the normal solute ions tend to occupy the d orbitals of the transition solvent. This effectively decreases ($E_O - E_F$) and the ATP, consequently, becomes larger and more negative. The more effective a given solute atom is in filling the d orbitals of the host atoms, the greater will be this effect. This also is shown in Table 7-3.

The effectiveness with which a normal metal atom can contribute electrons to affect E_F cannot always be based upon its valence in the ground state. This is particularly

Table 7-3
THERMOELECTRIC PROPERTIES OF BINARY ALLOYS

Base metal	Mg	Cu									Ag				Au								
Dissolved metal	Al	Al	Fe	Co	Ni	Au	Zn	Sn	Ag	Pd	Cu	Au	Pd	Al	Ag	Pd	Cu	Ti	Cr	Mn	Fe	Ni	Co
Effect on thermoelectric power[a]	−	−	--	--	--	−e	−	−	−e	−	−e	−e	−	−	−e	−	−e	--	--	--	--	--	--
Position in periodic table[b]	−	−	−	−	−	0	−	−	0	−	0	0	−	−	0	−	0	−	−	−	−	−	−

Base metal	Zn			Cd	Al											Sn	
Dissolved metal	Cu	Cd	Sn	Zn	Ag	Cu	Mg	Zn	Sn	Si	Ti	Cr	Mn	Fe	Ni	Pb	Zn
Effect on thermoelectric power[a]	+	−e	−	−e	+	+	+	+	−	−	+ +	−	--	--	−	0	+e
Position in periodic table[b]	+	0	−	0	+	+	+	+	−	−	−	−	−	−	−	0	+

Base metal	Fe								Ni							Co	Pd			Pt			
Dissolved metal	(C)	Mn	Ni	Cr	W	Mo	Si	Al	Cu	Cr	V	Mo	W	Mn	Al	Cu	Au	Ag	Cu	Pt	Ir	Rh	Pd
Effect on thermoelectric power[a]	--	--	--	+	+e	+e	--	−	−	+ +	+ +	+ +	+ +	−	--	+ +	−	−	−	+e	+	+ +	+e
Position in periodic table[b]	i.s.	+	−	+	+	+	+	+	+	+	+	+	+	+	+	+	+	+	+	0	+	+	0

[a] +, −: increase, decrease; + +, --: strong increase, decrease; +e, −e: slight increase, decrease.
[b] +: dissolved metal before base metal; −: dissolved metal after base metal; 0: metals in same group; i.s.: interstitial solutions.

After Crussard, C., in *Report of a Conference on Strength of Solids*, The Physical Society, London, 1948, 119. With permission.

true of normal metals in solid solutions in transition elements. Their s and p levels resonate between their own states and the vacant d states of the transition elements. Thus, on the average, only a fraction of these can be considered as occupying d levels.

For example, magnesium in the ground state has a valence of two. In contrast to this, when Mg is in solid solutions of this type, it appears to have a valence of about 0.6. This may vary somewhat, depending upon the host ion, and the kinds and quantities of other ions present in the solid solution. The actual valence of Mg, or any other atom, as a function of composition in any desired solid solution may be determined by measurements of the electrical resistivity and the application of Equations 6-24, 6-25, and 6-32.

This small valence has very important practical applications because Mg is frequently used to deoxidize melts of thermocouple alloys. A small excess of any deoxidant is always allowed to remain to ensure the complete removal of oxygen from the melt. The small residual amount of Mg in the solid solution has an almost negligible influence on the Fermi level because of its low valence; this effect frequently is neglected in practice.

Aluminum is another commonly used deoxidant. It has an effective valence of about 2.85; this is nearly five times as effective in the contribution of valence electrons as is Mg. Thus, a given residual amount of Al would have an appreciable effect upon E_F and, consequently, upon the ATP of the alloy; this must be taken into consideration if the thermocouple alloy is to produce an emf which will match the nationally accepted emf-temperature relationships within acceptable limits of error, especially in the higher temperature ranges.

7.12.2. Transition Metal Solutes

Transition metal solute ions may be considered as consisting of two categories. The first consists of those transition ions with relatively large numbers of available d orbitals, such as V, Cr, Mo, and W. In solution in other similar transition elements such as iron, they effectively increase the value of $(E_o - E_F)$ by sharing common s-d levels. Thus, they cause S to be smaller, or less negative. The detailed explanations of the anomalous behaviors of Cr, Mo, and W are beyond the scope of this text.

The other category of solutes is represented by such elements as Ni, Pt, and Pd whose d levels are more nearly filled. Again, the influence of transition alloying elements must be considered in relation to the degree to which the s and d levels of the host are shared between them. If the alloy ion has fewer d orbitals available than the solvent atom, it will decrease the value of $(E_O - E_F)$ and cause S to become larger, or more negative. If the solute atom has more orbitals than the solvent atom, the value of $(E_O - E_F)$ will increase. The ATP of such an alloy will become smaller, or less negative.

As mentioned in Section 7.9.4, when the electron concentration approaches that of the maximum filling of the d band, $(E_O - E_F)$ is no longer constant. When the electron concentration exceeds this maximum, increasing s-like behavior occurs, $(E_O - E_F)$ increases, and the magnitude of S decreases. Cu-Ni, Ag-Pd, and Au-Pd alloys behave in this way. The magnitude of S as a function of composition goes through a maximum at a given temperature.

Commercial constantans, complex Cu-Ni alloys, also show this behavior; this has important practical consequences (see Section 7.14.2).

7.13. APPLICATIONS TO PHASE EQUILIBRIA

Thermoelectric properties can provide a sensitive means for the determination of the limits of phase fields in constitution diagrams. Isothermal mappings of phase equilibria are obtained readily. This is accomplished by the use of suitable "families" of alloys. The principal variable in each such family is the quantity of alloying element for which the phase boundaries are to be determined. The discontinuities in a suitable series of isothermal plots define the phase boundaries in a way similar to that described in Section 6.5, Chapter 6.

For a given isotherm Equation 7-39, for normal metal solvents, becomes

$$S(T) = \frac{\text{constant}_1}{E_F} \tag{7-59}$$

and for transition metal solvents, Equation 7-49 becomes

$$S(T) = \frac{\text{constant}_2}{(E_O - E_F)} \tag{7-60}$$

Thus, at a given temperature the ATP of an alloy is a function only of its Fermi level.

Within a range of solid solubility the solute additions change the Fermi level and the ATP in a regular way. However, a discontinuity occurs in the Fermi level and the ATP as a function of composition when a second phase appears. This results from the fact that beyond the single-phase field, the chemical potential, or E_F, of each component no longer varies regularly with composition. Here, it is the same in each phase present. The chemical potential, or E_F, of a two-phase alloy follows the law of mixtures and is the weighted sum of the chemical potentials of the phases present. Under the conditions of constant pressure and constant temperature, the chemical potential is identical to the Fermi energy (Section 5.4.1, Chapter 5, Volume I). So, when a second phase appears or disappears, E_F will undergo a discontinuity which is reflected by the ATP.

It is apparent that a series of plots of ATP with such discontinuities, each determined at a different temperature, defines the phase boundaries of a given alloy system. The technique, obviously, is limited to temperatures below the solidus.

The above method provides valid results only if the conditions of the specimens approach equilibrium at the test isotherm. Supersaturation could occur if the specimens were cooled too quickly after annealing. On the other hand, subsaturation could occur if the specimens were heated too rapidly to the test temperature. Either or both of these conditions possibly might not be detected at low temperatures, or at any temperature at which the diffusion rate of the element under consideration is low. These could lead to erroneous results.

A typical determination of the limit of solid solubility by thermoelectric means is given in Figure 7-9.

Plots such as that given in the figure are used as shown in Figure 6-6. They have several important advantages over comparable determinations made by measurements of electrical resistivity. In the first place, thermoelectric measurements are much more sensitive to small variations in composition than are resistivity measurements. In addition, they are easier to perform and less susceptible to the introduction of errors; the length and cross-sectional area of a specimen do not enter into the calculations for ATP. The only requirements are that the temperatures of the reference and measuring junctions be known accurately. Both techniques require accurate chemical analyses of the specimens.

7.14. THERMOCOUPLE ALLOYS IN COMMON USE

Practically all of the thermocouple elements in common use are alloys of transition elements. These alloys have been selected by trial and error, over many years, because

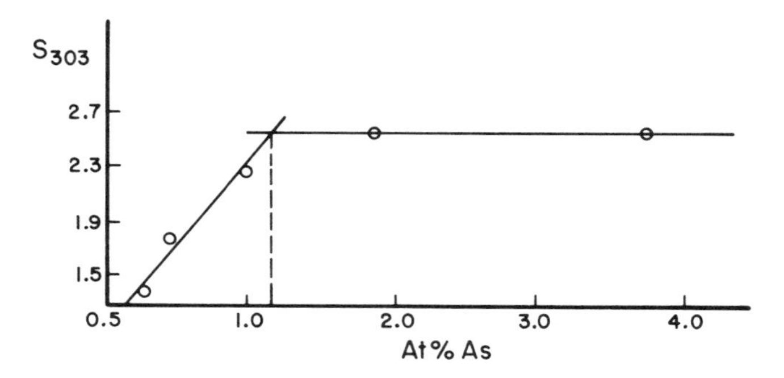

FIGURE 7-9. An example of the determination of the limit of solid solubility by thermoelectric means: As in Ag at 30°C.

Table 7-4
COMMON THERMOCOUPLES

I.S.A. Designation	Thermocouple	
	Positive leg	Negative leg
Type S	Pt — 10 Rh	Pt
Type R	Pt — 13 Rh	Pt
Type B	Pt — 30 Rh	Pt — 6 Rh
Type J	Iron	Constantan
Type T	Copper	Constantan
Type K	Chromel[a]	Alumel[a]
Type E	Chromel[a]	Constantan

Note: The trademarks and manufacturers of equally acceptable Type K thermocouple components are as follows. Type K positive leg: Tophel, Wilbur B. Driver Company; T-1, Driver-Harris Company; Thermokanthal KP, Kanthal Corporation. Type K negative leg: Nial, Wilbur B. Driver Company; T-2, Driver-Harris Company; Thermokanthal KN, Kanthal Corporation.

[a] Trade names — Hoskins Manufacturing Company.

they show nearly linear thermoelectric behavior as functions of temperature and because of the relatively large emfs that they generate (see section 7.8.7). Those most commonly used are given in Table 7-4. In addition, these thermocouples give stable and reproducible temperature readings when they are used properly. This results from the fact that all of the alloy thermoelements are solid solutions.

7.14.1. DILUTE ALLOY THERMOCOUPLE ELEMENTS

It is expected that this class of alloys should show linear thermoelectric properties as a function of temperature. The differentiation of Equation 7-51 gives

$$\frac{dS}{dT} = -\frac{\pi^2 k_B^2}{6e(E_o - E_F)} \tag{7-61}$$

It will be recalled (Section 7.9.3) that $(E_o - E_F)$ is virtually constant for these dilute

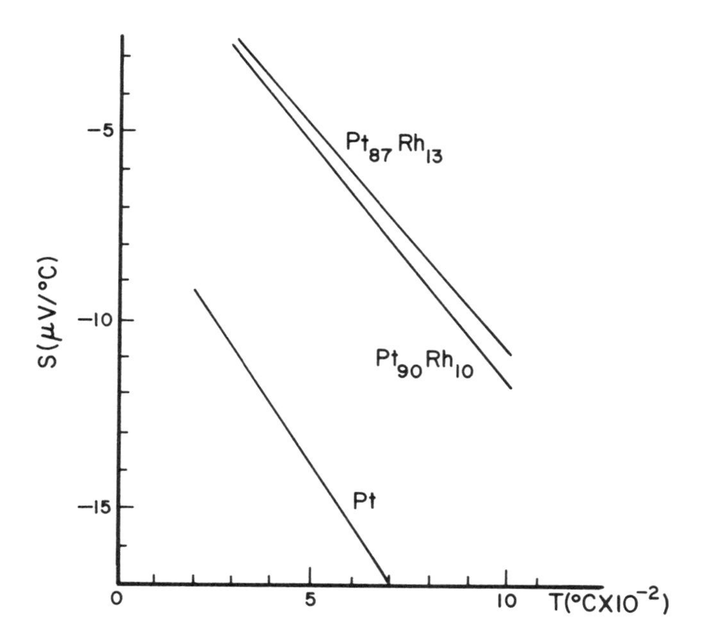

FIGURE 7-10. Thermoelectric properties of platinum and platinum-rhodium alloys.

alloys. Thus, Equation 7-61 is a constant for a given, dilute alloy and S is a linear function of T, with a negative slope. In this case Equation 7-43 must be considered. Here small changes in E_F result in negligible changes in $N(E_F)$. Thus $\ell n\ \phi(E)$, in Equation 7-41 is not a significant function of temperature and $(E_o - E_F)$ is virtually constant. Both the platinum-base and nickel-base alloys show this behavior.

The thermoelectric properties of Pt and its principle thermoelectric alloys are shown in Figure 7-10. It will be noted that the curves for these thermocouple elements are not quite parallel to that of platinum. This means that the emfs of the thermocouples they form with platinum are not exactly linear functions of temperature (see Section 7.7).

The addition of rhodium to platinum has the effect of decreasing the magnitude of S. The presence of the Rh ions in solution in the Pt has the effect of adding holes to the common d band, thus effectively increasing $(E_o - E_F)$. The positive displacement of the curves for the two alloys also is thought to result from the holes supplied by the Rh atoms. They are used at temperatures up to about 1540°C (2800°F).

These thermoelements should not be used in reducing atmospheres. Platinum and its dilute alloys, in such atmospheres as those containing hydrogen, carbon monoxide, and organic gases, will accelerate the decomposition of the normally refractory ceramic insulators used to separate the thermoelements. The metallic constituents of the decomposed ceramics react with the platinum and/or its alloys and usually form intermetallic compounds; some solid solutions also may be formed. In either event, the compositions of the thermoelements are changed permanently; their Fermi levels and, consequently, their ATPs change. This results in erroneous temperature readings. Chemical reactions of this kind may start as low as 600°C. The error in the thermocouple will begin to be evident at that point; it will increase as the reaction proceeds. Pt - Pt Rh thermocouples are most reliable when used in oxidizing atmospheres in which such degrading effects do not take place; they are highly resistant to chemical attack under these conditions.

The dilute alloys of nickel (Chromel and Alumel) show thermoelectric behaviors

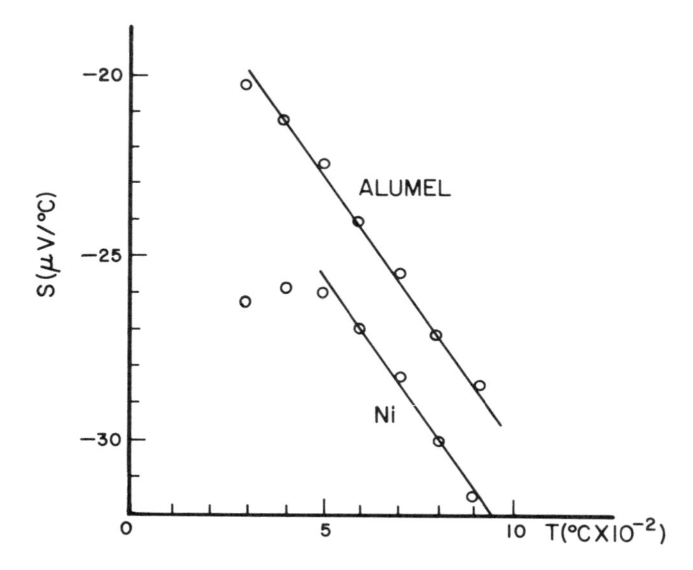

FIGURE 7-11. Absolute thermoelectric properties of nickel and Alumel.

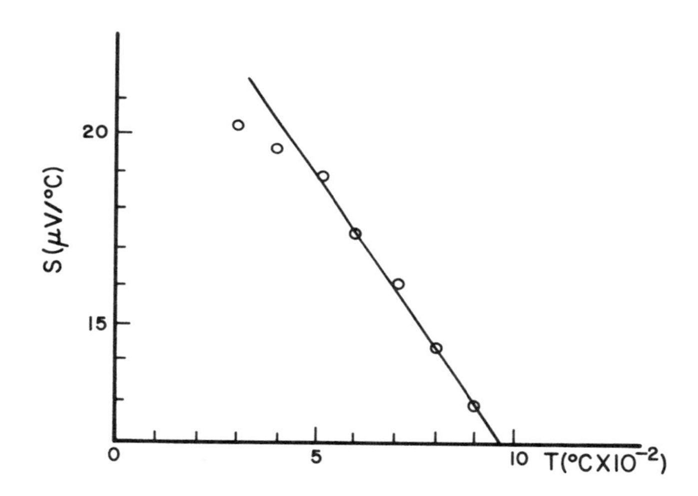

FIGURE 7-12. Absolute thermoelectric properties of Chromel.

similar to those of the platinum-base alloys. The thermoelectric data for nickel and Alumel are shown in Figure 7-11.

It will be noted that the behavior of nickel shows large deviations from linearity below 500°C. This results from the electron spin alignment effects which are responsible for the ferromagnetism in nickel. The Curie temperature of nickel is about 360°C. It is apparent that some short-range magnetic spin interactions are still present between 360 and 500°C. This type of electron behavior was not included in Equation 7-41, so Equation 7-51 cannot account for the observed results. At temperatures greater than 500°C the magnetic spin effects are absent, the metal is paramagnetic and Equation 7-51 is obeyed. In this range the thermoelectric power is large and negative and has a negative slope, as predicted by Equation 7-51.

At temperatures below 500°C, Alumel shows similar deviations, but to a lesser extent than does nickel. These are thought to arise from the same source as in the case

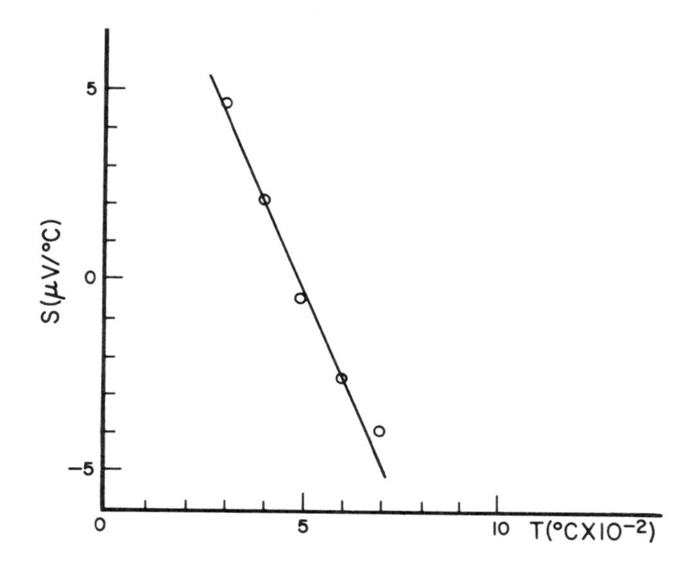

FIGURE 7-13. Absolute thermoelectric properties of thermocouple iron.

of nickel, but that the alloying elements (Mn, Si, and Aℓ) largely suppress this behavior; the ATP, therefore, conforms more closely to Equation 7-51.

Chromel shows behavior intermediate between that of nickel and Alumel with respect to deviations from linearity (Figure 7-12).

Chromel is anomalous in that it is positive in sign, probably as a result of the predominance of d-level hole conduction. It also shows deviations below 500°C. Its slope, however, is negative and close to that of nickel and Alumel. The nonlinear portions of the curves for Chromel and Alumel are roughly parallel. This minimizes the nonlinear effects below 500°C and permits their combination as a practical thermocouple up to about 1260°C (2300°F).

These Ni-base thermoelements also should not be used in reducing atmospheres or those containing even small amounts of sulfur. They maintain their stabilities in oxidizing atmospheres because their surfaces are protected by thin, continuous, adherent, protective, oxide films. These complex oxide films are destroyed in reducing atmospheres; the alloying elements than become subject to selective attack by the atmosphere, depleting the compositions of the alloys. The result is that ($E_O - E_F$) changes and the ATP of each of the thermoelements changes accordingly. This also can take place in oxidizing atmospheres when the partial pressure of the oxygen falls below a critical value. This behavior is particularly true of Chromel-type alloys.

The thermoelectric properties of thermocouple iron are intermediate between those of Chromel and nickel (Figure 7-13).

The sign of S is positive below about 500°C and negative above that temperature. This mixed behavior probably indicates a predominance of hole behavior at the lower temperatures with increasing amounts of conduction of electron character as the temperatures increase.

Thermocouple iron contains many alloying elements, all of which are present in very small quantities. Some of these are "tramp" elements which are picked up in the steelmaking process. All of the elements present have strong effects upon the ATP. The production of thermocouple iron is difficult because of the effects of the tramp elements. The number and quantity of these elements have continued to increase with time because of the scrap used in melting this material. Thermocouple iron (heavy

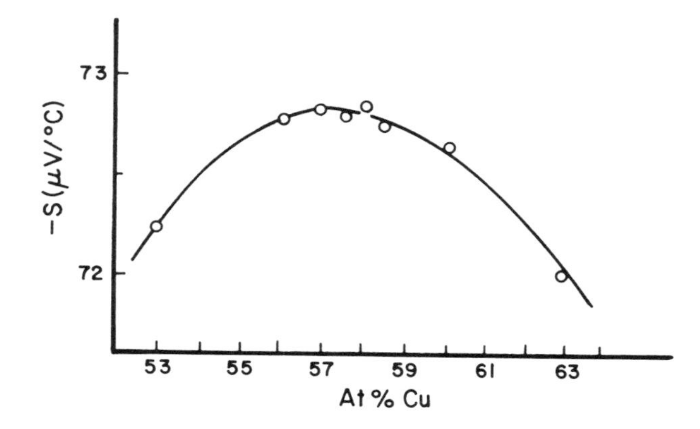

FIGURE 7-14. ATP of Cu-Ni alloys at 900°C.

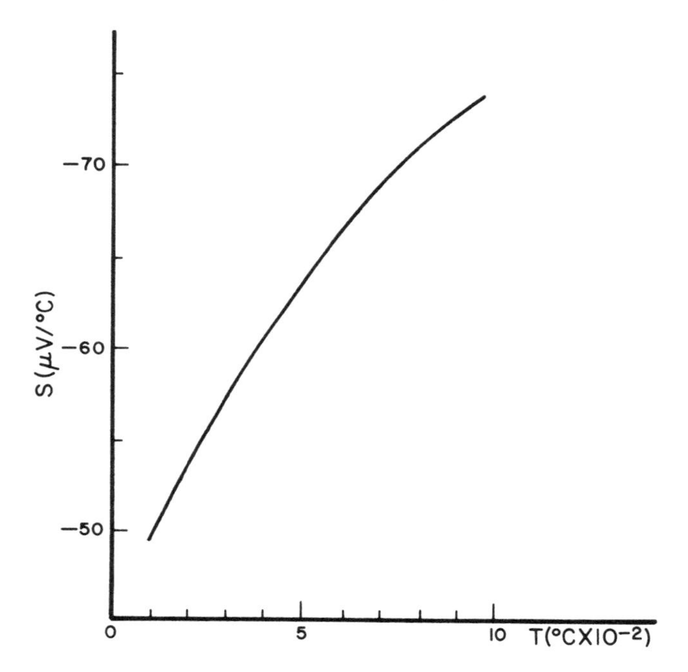

FIGURE 7-15. ATP of Cu-Ni alloys in the composition range from 56 to 59 at.% Cu.

gages) should not be used above about 870°C (1600°F). It may be used in both oxidizing and reducing atmospheres with little, or no, loss accuracy of measurement.

7.14.2. Concentrated Alloy Thermoelements

This class of alloys includes the Cu-Ni, Ag-Pd, and Au-Pd systems in which the noble metal is present at concentrations close to 60 at.%. Of these, the constantans (Cu-Ni) have been most widely used, and will be discussed here.

As noted in Section 7.9.4, these alloys represent the case in which maximum filling of the d band occurs. Here $(E_o - E_F)$ is quite small and is a function of temperature.

How does the band structure of a transition element, such as nickel, change when it is alloyed with a noble element, such as copper? The nickel atom lacks 0.6 electrons in its d levels. The copper atom has completed inner shells and one, outer, s electron.

The s electrons from the copper ions resonate to the d levels of the nickel ions until at 60 at.% these are maximumly filled.

This electron behavior is verified by the magnetic behavior of Cu-Ni alloys. The saturation magnetization of these alloys diminishes linearly with increasing copper and reaches zero close to 60 at.% Cu (see Figure 9-17, in Chapter 9). Here the ferromagnetic behavior vanishes. This results from the progressive filling of the d levels of the nickel ions by the s electrons from the copper ions until the d band is maximumly filled. Beyond this composition the alloys are paramagnetic.

As the filling of the d levels of nickel proceeds, $(E_o - E_F)$ becomes increasingly smaller and the magnitude of S increases until the d levels become maximumly filled. The factor $(E_o - E_F)$ approaches a minimum, nonzero value. At this point the probability of finding an added s electron in the d levels is quite small; that of finding it in the s band increases because of the s-d overlap. At electron concentrations greater than that of maximum filling, increasing s-like behavior is observed and $(E_o - E_F)$ becomes larger. As this occurs, S becomes smaller. Thus, a maximum in S occurs close to the 60/40 concentration at which the d levels of nickel become maximumly filled. This is shown in Figure 7-14.

It will be noted that this curve is quite flat in the range from 56 to 59 at.% Cu. Thus, the differences in the thermoelectric properties of alloys in this compositional range are quite small. This is shown in Figure 7-15.

The addition of alloying elements to a 57/43 Cu/Ni base has interesting effects. If the added element is such that its inner electron shells are completely filled, its valence electrons show s-like behavior in the bands of nickel and $(E_o - E_F)$ becomes larger. In this case S diminishes. If the added element is a transition element, holes are added and $(E_o - E_F)$ becomes larger. Thus, any alloying additions to the 57/43 Cu/Ni base decrease its magnitude of thermoelectric power. Commercial constantans contain varying amounts of such added elements. These elements are added to adjust the emf of the 57/43 Cu/Ni base so that it matches the accepted values and to enhance its resistance to high-temperature oxidation.

Constantans usually are combined with thermocouple irons (Type J). This combination may be used in both oxidizing and reducing atmospheres. The limiting temperature of application is determined by that of the iron leg. Type T thermocouples (copper-constantan) also may be used in both classes of atmospheres. The maximum temperature of application of these thermocouples is about 400°C (750°F). Above this temperature, components of the ambient atmospheres, especially oxygen, dissociate on the surface of the copper and either form solutions or compounds. These reactions change the emf of the thermocouples. Constantan also is paired with Chromel-type alloys (Type E). Here, the conditions of its use are controlled by the limits which must be placed upon the Chromels.

7.15. PROBLEMS

1. Show that the Peltier effect can be separated into the inherent properties of the components of a thermocouple.
2. The emf of a thermocouple is 27.5 mV at 500 K. Calculate the Peltier effect at the junction. Consider the units.
3. Calculate the entropy change in the junction of Problem 2.
4. The emf of a thermocouple is 3.899 mV at 873 K. The ATP of one thermoelement is $-11.66\ \mu V/°C$. Calculate the ATP of the other thermoelement.
5. Compare the merits of the following pairs of thermoelements for use in thermometry and cite their advantages and disadvantages: Cu-Ag, Cu-Au, Pt-Pd, and W-Mo.

6. Given that $S_A = 0.81 + 3.8 \times 10^{-3}T$ and $S_B = 0.28 + 5.5 \times 10^{-3}T$

   1. Find the expression for the thermoelectric power of such a thermocouple.
   2. Make a graph of the thermoelectric power of this couple as a function of temperature.
   3. Compute the emf of this thermocouple if the reference junction is at 100°C and the measuring junction is at 200°C.
   4. Discuss the suitability of such a thermocouple for practical use.

7. The Fermi energies (for free electrons) of Na and Cu are 3.1 and 7.0 eV, respectively. Calculate the ATP of these elements as a function of temperature.
8. Calculate the molar change in the free energy of Na and Cu as a function of temperature.
9. Estimate the electron contribution to the heat capacity of a transition element if $(E_o - E_F)$ is about 1 eV.
10. Estimate the change in E_F if a transition thermoelement showed a change of 13μV at 1100 K due to cold working.
11. Calculate the change in E_F of Ag when 2% of Aℓ is in solid solution in it. Assume Z Aℓ = 3.
12. Make an approximation of the rate of change of $N_d(E)$ as a function of $(E_O - E_F)$. Calculate this for the region $E_F \rightarrow E_O$. Make another series of calculations in the range $(E_O - E_F) \simeq 2$ eV for the first of these approximations.
13. If the constant in Equation 7-58 is $0.3\ (7 - Z_\beta)$ for Ag-base alloys, make an approximation for the ATP of binary alloys, each containing 1% of Zn and Ge, at 25°C.
14. Calculate the constants of Equation 6-24 from the data given for Mg and Aℓ in Section 7-11. Calculate the effects of 0.02 at.% of each of Mg and Aℓ on E_F and on ATP as binary constituents in solid solution in Cu at 400°C. What conclusions can be drawn from these data?
15. Determine the limits of solid solubility at 303 K of the following elements in Ag:

Aℓ (%)	S(303) (μV)	Ge (%)	S(303) (μV)	Sb (%)	S(303) (μV)
0.95	−1.09	0.35	−1.45	0.18	−1.42
1.88	−1.31	0.63	−1.82	0.23	−1.58
3.04	−1.37	1.23	−2.08	0.74	−2.52
3.71	−2.21	1.80	−2.13	2.08	−2.87

16. Use the emf data in the table given on the next page and in Table 7-2 to calculate the ATP of $Pt_{90}\ Rh_{10}$ and $Pt_{87}\ Rh_{13}$. Then determine the following:

   1. the constants of Equation 7-52 for each alloy
   2. the electron contribution to the heat capacity of each alloy
   3. the extent of the departures of the ATP of each alloy from linearity as compared to Pt

T(°C)	Pt_{90} Rh_{10}(mV)	Pt_{87} Rh_{13}(mV)
0	0	0
50	0.299	0.296
100	0.645	0.647
150	1.029	1.041
200	1.440	1.468
250	1.873	1.923
300	2.323	2.400
350	2.786	2.896
400	3.260	3.407
450	3.743	3.933
500	4.234	4.471
600	5.237	5.582
700	6.274	6.741
800	7.345	7.949
900	8.448	9.203
1000	9.585	10.503

7.16. REFERENCES

1. **Pollock, D. D.,** *The Theory and Properties of Thermocouple Elements,* S.T.P. 492, American Society for Testing and Materials, Philadelphia, Pa, 1971.
2. **Dike, P. H.,** *Thermoelectric Thermometry,* Leeds and Northrup, Philadelphia, Pa., 1954.
3. **Benedict, R. P.,** *Manual on the Use of Thermocouples in Temperature Measurement, 2nd ed.,* S.T.P. 470A, American Society for Testing and Materials, Philadelphia, Pa., 1974.
4. **Mott, N. F. and Jones, H.,** *The Theory of the Properties of Metals and Alloys,* Dover, New York, 1958.
5. **Wilson, A. H.,** *Theory of Metals,* 2nd ed., Cambridge, New York, 1953.
6. **Finch, D. I.,** *Principles of Thermoelectric Thermometry,* Leeds and Northrup, Philadelphia, Pa, 1962.
7. **Hertzberg, G., Ed.,** *Temperature, Its Measurement and Control in Science and Industry,* Vols. I, II, and III, Reinhold, 1962.
8. **Bridgman, P. W.,** *Thermodynamics of Electrical Phenomena in Metals,* Dover, New York, 1960.
9. **MacDonald, D. K.,** *Thermoelectricity: An Introduction to the Principles,* John Wiley & Sons, New York, 1962.
10. **Hume-Rothery, W.,** in *Electronic Structure and Alloy Chemistry of the Transition Elements,* Beck, P. A., Ed., John Wiley & Sons, New York, 1963.
11. **Crussard, C.,** in *Report of a Conference on Strength of Solids,* The Physical Society, London, 1948.
12. **Wang, T. P., Starr, C. D., and Brown, N.,** *Acta Met.,* 14, 649, 1966.
13. **Swindells, J. F., Ed.,** Precision Measurement and Calibration, Special Publ. No. 300, Vol. 2, National Bureau of Standards, U.S. Government Printing Office, Washington, D.C., 1968.

Chapter 8

DIAMAGNETIC AND PARAMAGNETIC EFFECTS

All matter is affected by magnetic fields; there are no nonmagnetic materials. However, the magnetic response may be very small in many cases. One way to describe the behaviors of materials in magnetic fields is to categorize them on the bases of the magnitude, sign, and effect of temperature of their magnetic properties. This leads to classifications such as diamagnetic, paramagnetic, ferromagnetic, ferrimagnetic, antiferromagnetic, and other types of materials. The first two of these types are discussed in this chapter; the others are described in Chapter 9.

Many materials show a very small magnetic susceptibility. This may be composed of a diamagnetic part and a paramagnetic part. The diamagnetic component arises from the translational motion of the electrons of the material in the applied field. The paramagnetic contribution originates from the alignment with the applied field of those ions which have permanent magnetic moments. These permanent ionic magnetic moments result from angular and orbital spin properties of the electrons. A material will be either paramagnetic or diamagnetic depending upon which of the effects is predominant. In the case where the permanent magnetic moments are large, ferromagnetism can result.

Both of these effects may be small compared to the applied field. In this case, such materials frequently are designated incorrectly as being nonmagnetic.

The concepts developed and discussed in the subsequent sections for paramagnetism and diamagnetism provide a basis for additional understanding of electron behavior in solids. The ideas presented to explain paramagnetism are also employed in the explanation of ferromagnetic effects, since both are based upon the presence of permanent electric dipoles.

8.1. CLASSICAL BASIS FOR DIAMAGNETISM AND PARAMAGNETISM

A simple way to discriminate between diamagnetic and paramagnetic materials is to place them in a nonuniform magnetic field. A material that tends to move toward the region of lower field strength is diamagnetic. Conversely, if it moves in the direction of higher field strength, it is paramagnetic.

It is helpful to examine some elementary magnetic behavior and some of the units involved. A current flowing in a long, thin solenoid will produce a uniform magnetic field in its interior. A magnetic field is the volume, or three-dimensional space, in which a magnetic pole is affected by a magnetic force. Current, I, flowing in such a solenoid will give a field intensity of

$$H = nI \tag{8-1}$$

The current is in amperes and n is the number of turns of wire per meter. This defines the unit of H as amperes/meter. H is related to the oersted by H = 1 A/m = $4\pi \times 10^{-3}$ Oe, or 1 Oe = 79.6 A/m.

The attraction between two magnetic poles, of strengths p_1 and p_2, is an important means of exerting force. The force, F, between the poles, separated by a distance, d, is

$$F = \frac{P_1 P_2}{4 \pi \mu_o d^2} \tag{8-2}$$

The force is in newtons, N, the pole strengths are in webers. The unit of the weber is $1/4\,\pi \times 10^{10}$ gauss-cm^3/meter. Other relationships involving the weber are useful:

$$1\text{Wb/meter}^2 = 1\text{ N/A-meter} = 1\text{ tesla} = 10^4\text{ gauss}$$

The factor μ_o is the permeability of free space (vacuum), and is discussed below. Its units are henrys/meter (the henry is defined as being equal to 1 volt-sec/A). This factor is given by

$$\mu_o = 4\,\pi \times 10^{-7}\text{ H/m} \tag{8-3}$$

The force acting on a single pole, p, in a field, H, is

$$F = pH \tag{8-4}$$

When a magnetic dipole, or, e.g., a bar magnet, in which the poles are separated by a distance, d, is acted upon by a uniform field, H, a force couple is formed (see Figure 8-1). The moment, M_c, of the couple is given by

$$M_c = -pHd\sin\theta \tag{8-5}$$

Since H is uniform, the dipole only will rotate until it becomes parallel with the field. This is typical of paramagnetic behavior. Diamagnetic materials tend to align themselves across the magnetic field. No other motion can occur unless the field is nonuniform. For example, any motion of the dipole in the figure in the x direction, after having rotated parallel to the field, requires a gradient in the field in that direction. The motion results from the force

$$F_x = pd\frac{\partial H_x}{\partial x} \tag{8-6}$$

These two illustrations are given to show that the forces acting upon the dipole must involve the moment pd. Its units are the weber-meter. In the above illustration the field-induced torque on the dipole is

$$M_c = -MH\sin\theta \tag{8-7}$$

and the potential energy is

$$E(\theta) = -MH\cos\theta \tag{8-8}$$

under ideal, frictionless, conditions.

A magnetic moment per unit volume, M, is induced in a material when it is placed in a magnetic field, H. The sign of M indicates which of the two types of magnetism under discussion here is present. When M is negative, the material is diamagnetic; M is positive for paramagnetic substances. The vectors M and H are parallel and proportional to each other giving

$$M = \chi H \tag{8-9}$$

The constant of proportionality, χ, is the magnetic susceptibility. The molar susceptibility may be obtained when the unit volume ascribed to M corresponds to that of 1

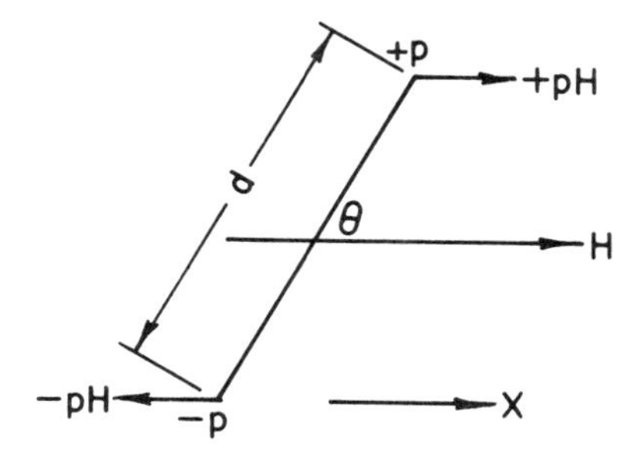

FIGURE 8-1. Magnetic dipole in a uniform magnetic field.

mol wt (1g-mol) of a substance. The mass susceptibility is given by the volume susceptibility divided by the density.

The induced field, or magnetic induction, B, is defined as

$$B = H + 4\pi M = \mu H \quad (8\text{-}10)$$

The unit of B is the henry/meter. The permeability, μ, is the ratio of the induced field to the applied field. It is a constant for paramagnetic and diamagnetic materials up to the point at which they approach saturation. Hysteresis effects in ferromagnetic materials (Chapter 9) result in a more complex relationship between B and H.

Equation 8-10 can be reexpressed as

$$H - \mu H = -4\pi M \quad (8\text{-}11)$$

or

$$H(1 - \mu) = -4\pi M$$

from which

$$1 - \mu = -\frac{4\pi M}{H}$$

Now, using Equation 8-9, this becomes

$$\mu = 1 + 4\pi\chi \quad (8\text{-}12)$$

Both B and H are vectors which measure properties of the same field and are related by Equation 8-10. The magnetic moment, M, also is a vector which has the same dimensions (Gaussian) as H. M may be defined as

$$M = \frac{B}{\mu_o} - H \quad (8\text{-}13)$$

where μ_o is the permeability of free space. When both sides of Equation 8-13 are divided by H,

$$\frac{M}{H} = \frac{B}{H\mu_o} - 1$$

If μ is the permeability of a substance, and is equal to B/H, Equation 8-10, then

$$\frac{M}{H} = \frac{\mu}{\mu_o} - 1$$

and the relative permeability of the material is

$$\mu_r = \frac{\mu}{\mu_o} = 1 + \frac{M}{H} \tag{8-14}$$

The magnetic susceptibility, χ, of Equation 8-9, is a constant for diamagnetic and paramagnetic substances. The substitution of this into Equation 8-14 gives

$$\mu_r = \frac{\mu}{\mu_o} = 1 + \chi \tag{8-15}$$

Diamagnetic substances have negative values of χ; those of paramagnetic materials are positive. Thus, on the basis of relative permeability, diamagnetic materials have $\mu_r < 1$, while paramagnetic materials have $\mu_r > 1$.

8.2. DIAMAGNETISM

Diamagnetism arises from the changes of the "orbital" axes of the electrons of an ion in an applied magnetic field. The current which is induced in the substance is opposite to the field (Lenz's law). Consider an electron orbiting about the nucleus of a Bohr-type atom as a resistanceless loop. The induced current will exist while the field is applied. This induced current produces a magnetic field which is opposite in direction to the applied field. The magnetic moment caused by the induced current is the diamagnetic moment. All substances show a diamagnetic component. However, this may be very small in comparison to such other phenomena as paramagnetic or ferromagnetic effects which also may be present.

Consider an orbit around a nucleus of average radius, r, containing Z electrons, each with a charge e, as comprising a resistanceless loop. The time for the charge to make one circuit around the loop is $2\pi/\omega_o$, where ω_o is the angular velocity. This permits the calculation of the current in the loop:

$$\text{Current} = I = \frac{\text{charge}}{\text{time}} = Ze \div \frac{2\pi}{\omega_o} = \frac{Ze\omega_o}{2\pi} \tag{8-16}$$

From electricity theory the magnetic dipole moment is, in Gaussian units, where A is the average area of the loop and c is the velocity of light,

$$\mu_M = \frac{IA}{c} = -\frac{1}{c} \cdot \frac{Ze\omega_o}{2\pi} \cdot \pi r^2$$

or

$$\mu_M = \frac{Ze\omega_o r^2}{2c} \tag{8-17}$$

This is perpendicular to the plane of the loop. The angular momentum of an electron is given by Equation 1-26 as $mr^2{}_{\omega_o}$. This is used in Equation 8-17 to obtain

$$\mu_M = Z\mu_e = Z\left[\frac{e}{2mc}\right] p(\theta) \tag{8-18}$$

where the quantity within the brackets is the magnetic moment of a single electron. In diamagnetic behavior, Equation 8-17 and, therefore, Equation 8-18 are negative because the dipole moment is vectorially opposite to the angular momentum of the electron, $p(\theta)$.

The force acting on an electron is

$$F = m\omega_o^2 r \qquad (8\text{-}19)$$

When a field is applied normal to the plane of the electron loop, the force becomes

$$F = m\omega_o^2 r - F_N \qquad (8\text{-}20)$$

where F_N is the force resulting from the applied field. The normal force is given by

$$F_N = \frac{e}{c}(\vec{v} \times \vec{H}) = \frac{e}{c} vH \qquad (8\text{-}21a)$$

Then, recalling that the velocity $v = \omega r$, where ω is the angular velocity of the electron in the field, Equation 8-21a becomes

$$F_N = \frac{e}{c}\omega rH \qquad (8\text{-}21b)$$

Equation 8-21b is substituted into Equation 8-20 to give the net force as

$$F = m\omega_o^2 r - \frac{e}{c}\omega rH \qquad (8\text{-}22)$$

Now substituting for F in Equation 8-22, in terms of the changed angular velocity, gives

$$F = m\omega^2 r = m\omega_o^2 r - \frac{e}{c}\omega rH$$

and rearrangement results in

$$mr(\omega^2 - \omega_o^2) = -\frac{e}{c}\omega rH \qquad (8\text{-}23)$$

This also can be rewritten as

$$\omega^2 - \omega_o^2 = -\frac{e}{c}\omega rH \cdot \frac{1}{mr} = -\frac{e\omega H}{mc} \qquad (8\text{-}24)$$

The quantity $(\omega^2 - \omega_o^2)$ can be factored and Equation 8-24 is rearranged to give

$$\omega - \omega_o = -\frac{eH}{mc} \cdot \frac{\omega}{\omega + \omega_o} \qquad (8\text{-}25)$$

The approximation can be made that $\omega \simeq \omega_o$ because the applied fields are capable of inducing only small changes in ω_o. This results in

$$\Delta\omega = \omega - \omega_o = -\frac{eH}{2mc} = \omega_L \qquad (8\text{-}26)$$

in which ω_L is the Larmor precession frequency (Figure 8-2a). In other words, the application of the field causes the electrons to precess in their orbits because the external field exerts a torque on the electron and the orbit precesses with a frequency of eH/2mc (see Equations 8-76b and 8-77b.)

The expression for the magnetic moment (Equation 8-17) can be differentiated to obtain

$$\Delta\mu_M = \frac{Zer^2\Delta\omega}{2c} \tag{8-27}$$

Equation 8-26 is substituted for $\Delta\omega$ to give

$$\Delta\mu_M = -\frac{Zer^2}{2c}\cdot\frac{eH}{2mc} = -\frac{Ze^2r^2H}{4mc^2} \tag{8-28}$$

When this fraction is multiplied by ω/ω, Equation 8-28 can be rewritten as

$$\Delta\mu_M = -\frac{Zer^2\omega}{2c}\cdot\frac{1}{\omega}\cdot\frac{eH}{2mc}$$

Now, again using Equation 8-17, this becomes

$$\Delta\mu_M = -\frac{\mu_M}{\omega}\cdot\frac{eH}{2mc} \tag{8-29}$$

in which the second factor is given by Equation 8-26. Upon substitution, Equation 8-29 becomes

$$\Delta\mu_M = \frac{\mu_M}{\omega}(\omega_L) = \frac{\omega_L}{\omega}\mu_M \tag{8-30}$$

If a solid consists of N such ions, each of which has more than one electron in several "orbits" of radii r_i, then, by Equation 8-28,

$$M = N\Delta\mu_M = -\frac{NZe^2}{4mc^2}H\sum_i r_i^2 \tag{8-31}$$

This gives the diamagnetic susceptibility, using Equation 8-9, as

$$\chi_D = \frac{M}{H} = \frac{N\Delta\mu_M}{H} = -\frac{NZe^2}{4mc^2}\sum_i r_i^2 \tag{8-32}$$

in which the summation is made for the electrons of one ion of the solid.

Thus, the diamagnetic susceptibility arises from the small changes in angular velocity which result in the Larmor precession of electrons which are induced by the applied field (Figure 8-2). The electrons responsible for this are usually found in the filled shells. This accounts for the presence of diamagnetic components in all materials to some degree.

As H increases from zero (Equation 8-28) there is no change in the shape or size of the electron orbit. However, the orbit starts to precess such that its axis, or pole, (orbit normal) sweeps out a cone about H. Since the electron carries a charge, this precession induces a magnetic moment opposite to that of H. This is shown in Figure 8-2b.

The magnetic polarization within a substance may be given more simply than that of Equation 8-31 by the product of the number of ions per unit volume and the change in the magnetic moment of an ion from Equation 8-28:

$$M = N\Delta\mu_M = -\frac{NZe^2r^2H}{4mc^2} \tag{8-31a}$$

The diamagnetic susceptibility, using Equation 8-9, is given by

$$\chi_D = \frac{M}{H} = -\frac{NZe^2r^2}{4mc^2} \tag{8-32a}$$

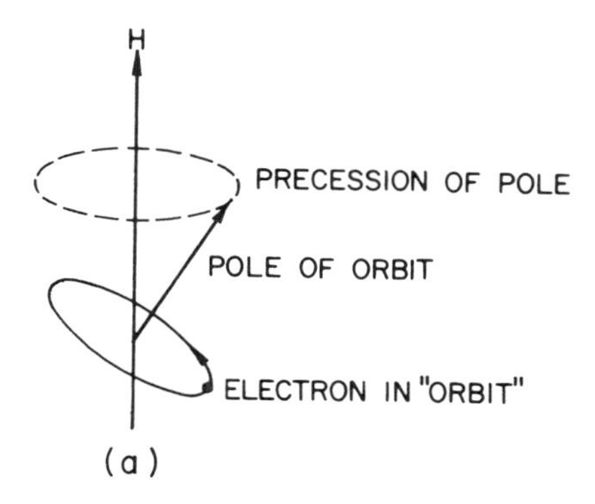

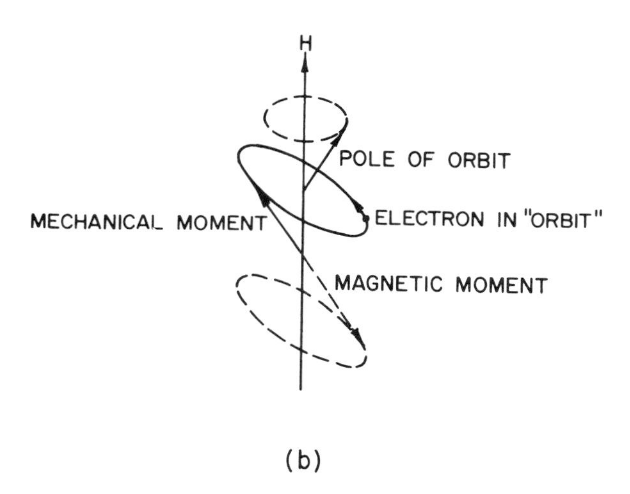

FIGURE 8-2. (a) Larmor precession. Pole precesses with frequency ω_L; (b) diagram showing mechanical and magnetic moments.

This derivation is based upon the assumption of a uniform, spherical charge distribution of a single electron shell of average radius, r. Where several electron shells are involved, the mean-square radius $\overline{r}^2$, is used such that

$$\overline{r}^2 = \frac{2}{3} r^2$$

Under this condition, Equation 8-32a gives the Langevin equations for magnetization and diamagnetic susceptibility:

$$M = -\frac{NZe^2\overline{r}^2H}{6mc^2} \quad \text{and} \quad \chi_D = -\frac{NZe^2\overline{r}^2}{6mc^2} \qquad (8\text{-}33)$$

It will be noted that Equations 8-32 and 8-33 give susceptibilities which are independent of both temperature and the applied field.

Implicit in the foregoing are the ideas that the direction of the applied magnetic field is parallel to an axis of symmetry of an isotropic crystal in the field, because the theory is most applicable to monatomic gases. Where the crystal is real and anisotropic, or is made up of molecules, this may not necessarily be the case. When these factors must be considered, the susceptibility as given by Equations 8-32, 8-32a, or 8-33 must be

Table 8-1
MASS SUSCEPTIBILITIES OF SOME OF THE ELEMENTS NEAR ROOM TEMPERATURE

Element	$\chi \times 10^6$	Element	$\chi \times 10^6$	Element	$\chi \times 10^6$
H	−1.97	Si	−0.13	Mo	+0.04
He	−0.47	P	−0.90	Pd	+5.4
Li	+0.50	Ca	+1.10	Ag	−0.20
Be	−1.00	Ti	+1.25	Cd	−0.18
B	−0.69	V	+1.4	Sn	−0.25
C	−0.49	Cr	+3.08	Sb	−0.87
N	−0.80	Mn	+11.8	La	+1.04
O	+106.2	Cu	−0.86	Ta	+0.93
Na	+0.51	Zn	−1.57	W	+0.28
Mg	+0.55	Zr	−0.45	Pt	+1.10
Al	+0.65	Nb	+1.5	Bi	−1.35

From Stanley, J. K., *Electrical and Magnetic Properties of Metals*, American Society for Metals, Metals Park, Ohio, 1963, 212. With permission.

modified by the inclusion of an additional positive term which results from polarized dipole moments. This term is called the Van Vleck paramagnetism. The substance will be either diamagnetic or paramagnetic depending upon the relative sizes of the two terms.

The mass susceptibilities of some of the elements are given in Table 8-1. It will be noted that χ_D is quite small, being of the order of 10^{-6}.

8.2.1. Crystalline Diamagnetic Materials

Equations 8-32 and 8-33 are applicable to freely rotating ions. However, other factors enter where crystals are involved. Most crystals are anisotropic so the environment about each ion varies according to the crystallographic direction. Under these conditions, the Larmor precession behavior described in Section 8.2 is not strictly applicable. However, a descriptive understanding of the diamagnetic behavior of materials can be obtained.

Diamagnetic effects may be overshadowed by paramagnetic effects. This is true in solids in which the electrons can make transitions, as a result of the influence of an applied field, to other quantum states with different magnetic moments (see Section 8.3.2). However, transitions of this kind usually are repressed when the orbital and spin moments of strongly bound, paired electrons cancel each other. This is the situation where atoms or molecules have completed electron levels. Electrons characteristically tend to form closed shells, especially in compound formation. This accounts for the fact that most solids are diamagnetic. Rock salt serves as a good illustration of this. The individual component atoms, Na and Cℓ, have magnetic dipole moments. In the solid NaCℓ crystal, the Na atom gives up its valence electron to the Cℓ atom. The ions in the resultant solid have closed-shell electron configurations like those of neon and argon, respectively. Under these conditions, the ions have no dipole moments and the substance is diamagnetic because of the reaction of the electrons in the closed shells to the applied field.

Covalently bonded solids behave in the same way. This results from the sharing of electrons between ions which effectively completes the electron shells. The resultant configuration has no magnetic dipole moment and the solid is diamagnetic.

Most chemical bonds are of mixed character, being partly ionic and partly covalent (see Section 10.6.3 in Volume III). Here, the wave functions of the bonding electrons

Table 8-2
EXPERIMENTAL VALUES OF DIAMAGNETIC SUSCEPTIBILITIES PER GRAMION FOR NOBLE GASES AND FOR IONS WITH NOBLE GAS ELECTRON CONFIGURATIONS

(Units of 10^{-6} EMU/mole)

He	Ne	A	Kr	Xe	F^-	$C\ell^-$	Br^-	I^-
1.9	7.6	19	29	44	9.4	24.2	34.5	50.6
Li^+	Na^+	K^+	Rb	Cs	Mg^{++}	Ca^{++}	Sr^{++}	Ba^{++}
0.7	6.1	14.6	22.0	35.1	4.3	10.7	18.0	29.0

Adapted from Myers, W. R., *Rev. Mod. Phys.*, 24, 15, 1952. With permission.

are not equally divided between the ions concerned, nor are they localized. Thus, where the bonds are of mixed character, the electrons can impart a very small paramagnetic moment. Under these conditions, diamagnetic behavior usually predominates.

From the above, it is readily seen why diamagnetism is present to some degree in all materials.

8.2.1.1. Ionic Crystals

The bonding in ionic crystals, such as the alkali halides, is virtually completely ionic. The ions have closed-shell configurations of the noble gases adjacent to them in the Periodic Table. To the extent that each ion in the $CsC\ell$, $NaC\ell$, and ZnS lattices has unlike nearest neighbors, (see Figure 10-13, Chapter 10, Volume III), these crystal types can be considered, to a first approximation, as being nearly isotropic. A similar approximation can be made for crystals composed of doubly ionized components. As an example, MgF_2 can be considered as being composed of ions with closed-shell electron configurations similar to that of neon.

The diamagnetic properties of substances such as these, where the bonding is almost completely ionic, may be understood on the same basis as that of individual ions where the Larmor precession is induced in each ion. This might lead to the expectation that their ionic susceptibilities could be determined from susceptibility data from a given family of compounds. Such a determination could be made if the susceptibility of one of the ions was known and if that of the other ion was constant in the given family of compounds. This might be done in the following way. Supposing that the susceptibility of Na^+ was known, the susceptibility of F^- could then be determined from that of NaF. Knowing the susceptibility of F^-, the susceptibility of Li^+ could be determined from measurements of LiF. It turns out that this approach leads only to approximate values of the susceptibilities of these compounds. The susceptibility of a given ion is not constant, but varies from compound to compound. The deviations are most pronounced in the salts with the smaller molecular weights.

Calculations of the susceptibilities of halides of the alkaline earths deviate more widely than those for the salts of the alkaline metals. These variations arise from the fact that overlap (see Section 10.6.3, Chapter 10, Volume III) occurs, distorting the ion. The Larmor precession is affected somewhat by the ionic distortion.

Some values of diamagnetic susceptibilities of ions are given in Table 8-2.

8.2.1.2. Covalent Crystals

The covalent bonds joining organic molecules are formed by the sharing of electrons by pairs of atoms, where the wave functions include both atoms. This poses difficulties to the understanding of diamagnetism since the details of the wave functions seldom are known. In addition, Equation 8-33, derived for a single atom, does not always hold accurately when several different atoms are involved.

However, close approximations of organic molecules can be made by adding the susceptibilities of the component atoms, groups of atoms, and electron bonds. An example given by Martin is quoted here. The molar susceptibility of ethyl bromide, C_2H_5Br, is calculated from the susceptibilities of its constituents (in EMU units $\times 10^{-6}$): C = −6.00, H = −2.93, Br = −30.6 and the C − Br bond = +4.1. Here, the C − Br bond forms a dipole and is paramagnetic. The calculation is as follows: $(-2 \times 6.00 - 5 \times 2.93 - 30.6 + 4.1) \times 10^{-6} = -53.1 \times 10^{-6}$ EMU. This is in excellent agreement with the experimental value of -53.3×10^{-6} EMU. Based on theoretical considerations, it has been shown that the susceptibility of such organic molecules must involve the sum of the effects of the component atoms. As noted in Section 8.2, the Van Vleck effect is also considered in order to make accurate calculations.

Bulk specimens of amorphous or polycrystalline organic materials will behave nearly isotropically because of the random orientations of their molecules or grains. Where a single crystal is made up of organic molecules, the diamagnetic properties of the crystal may be anisotropic because of the anisotropic bonding of the crystalline array. This is especially true of organic compounds which contain hexagonal "rings" and similar configurations. In many of these rings, the carbon atom has only three nearest neighbors instead of four, as might be expected (see Section 10.6, Chapter 10, Volume III). Thus, the usual covalent bonding is not present. The extra electrons occupy states which permit them to be present in any of the bonds of a ring, and to be moved easily. When a field is applied perpendicular to the ring, these readily moveable, extra electrons circulate around the rings. Their contribution to the diamagnetic effect is approached by Equations 8-32 and 8-33. However, in this case the radius is not that of the orbit, but that of the ring. This accounts for the relatively high susceptibility under these conditions. When the field is applied in the plane of the ring, it does not cause the extra electrons to circulate, in agreement with the theory. Thus, crystals of organic materials of this type are expected to show diamagnetic anisotropy.

A similar bonding array exists in graphite where each carbon atom in the basal plane has three nearest neighbors. The extra electron moves with relative ease in this plane and imparts high diamagnetic susceptibility. Here, it appears that the radius of these extra electrons must be approximately four ring diameters to account for this. This effect is reduced by phonons induced by elevated temperatures which distort the lattice. The anisotropy of graphite also is reduced when it is in the finely divided form, diminishing this effect.

8.2.1.3. Metals

The metallic lattice is one in which the valence electrons are nearly free and are shared by all of the ions in the lattice; they move readily throughout the lattice. The application of a magnetic field will affect the motion of these electrons and, consequently, magnetic effects are anticipated in addition to the diamagnetic effects of the filled shells of the ion cores.

The magnetic properties arise from both the translational motion and from the angular momentum associated with spin. The diamagnetic effects are a result of the translational motion. Paramagnetism is caused by spin and angular orbital momentum.

The application of a magnetic field will cause a "free" electron to move in a path along a helix whose axis is parallel to the applied field. This path results from the normal force given in Equation 8-21a. The projection of one turn on the helix onto a plane normal to the axis of the spiral gives a circle. The radius of the projected circle is obtained by equating the normal force (Equation 8-21a) and the centrifugal force acting upon the electron:

$$\frac{evH}{c} = \frac{mv^2}{r}$$

From which

$$r = \frac{mcv}{eH} \quad \text{and} \quad v = \frac{eHr}{mc}$$

so,

$$\omega = \frac{v}{r} = \frac{eH}{mc}$$

where v is the velocity, c is the speed of light, and ω is the angular frequency, and is known as the cyclotron resonance frequency. It is twice the Larmor precision frequency (Equation 8-26). Also see Equation 8-77b.

The quantum-mechanic treatment gives the energy in terms of the quantization of the angular velocity. This is obtained by considering the motion of a particle on the plane of projection as simple harmonic oscillation. Using Equation 3-72, in terms of ω, results in

$$E_n = (n + 1/2)h\nu = (n + 1/2)\frac{h}{2\pi}\omega + E_z$$

$$= \frac{(2n+1)}{2}\cdot\frac{h}{2\pi}\cdot\frac{eH}{mc} + E_z = E_{xy} + E_z$$

where the first term, E_{xy}, is the energy component perpendicular to the field, because of the projection of the spiral onto the plane normal to its axis, and E_z is the kinetic energy component parallel to the field and is unaffected by it. The Bohr magneton, μ_B, is defined as eh/2mc so, neglecting E_z,

$$E_n = (2n + 1)\left|\mu_B\right|H$$

Thus, in the presence of a magnetic field, the valence electrons, which formerly occupied states in a quasi-continuum, now reside in relatively widely separated, highly degenerate, discrete states known as "Landau levels". The forbidden energy ranges, or gaps, between these levels are given, just as in Section 3.10 and shown in Fig. 3-8 (Chapter 3, Volume I),

$$E_g = h\nu = \frac{h}{2\pi}\cdot\frac{eH}{mc} = 2\mu_B H$$

For example, electrons, in the absence of a field, which normally occupy states in the range $[2(n+1)+1]|\mu_B|H-(2n+1)|\mu_B|H = 2|\mu_b|H$ become degenerate and occupy a single state when a field is applied as shown in Figure 8-3.

It can be shown that this degeneracy is consistent with the principle of exclusion. When spin is expressly included in Eq. (5-19), it becomes

$$N(p) = \frac{8\pi Vp^2dp}{h^3}$$

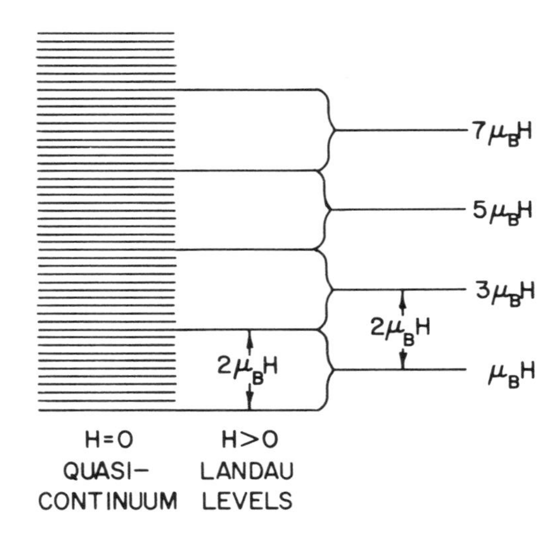

FIGURE 8-3. Landau levels resulting from the degeneracy of valence electron states in a magnetic field.

The number of momentum states at p in an interval dp is given by

$$\frac{N(p)}{p} = \frac{8\pi V p dp}{h^3}$$

This can be reexpressed, using $dp = dp_{xy} \cdot dp_z$, as

$$\frac{N(p)}{p} = \frac{8\pi V p_{xy} dp_{xy} dp_z}{h^3}$$

It will be recognized that $p_{xy} = 1/2d(p_{xy})^2 = mdE \simeq m\Delta E = 2m|\mu_B|H$. So,

$$\frac{N(p)}{p} = \frac{16\pi V m |\mu_B| H dp_z}{h^3}$$

This may be simplified by substituting for μ_B to get

$$\frac{N(p)}{p} = \frac{16\pi V m \frac{h}{2\pi} \frac{e}{2mc} H dp_z}{h^3} = \frac{4VeHdp_z}{ch^2}$$

Differentiating Equation 5-56 gives

$$dp_z = \frac{h}{2\pi} d\bar{k}_z$$

the substitution of which results in

$$\frac{N(p)}{p} = \frac{4VeH}{ch^2} \cdot \frac{h}{2\pi} d\bar{k}_z = \frac{2VeH}{\pi ch} d\bar{k}_z$$

This gives the degeneracy in a Landau state. The exclusion principle is not violated as long as the wave vectors of the electrons have different directions. In other words, exclusion is maintained as long as the paths of the electrons in a given Landau level are different.

The diamagnetic component of the valence electrons may be approximated from the behavior shown in Figure 8-3, along with reference to Figure 5-5c, Chapter 5, Volume I. The number of electrons in the range $E \simeq E_F$ is

$$N(E_F) = f(E)N(E,0) = \frac{1}{2} N(E_F, 0)$$

The total number of electrons is approximated by

$$N \simeq N(E_F, 0)E_F$$

$$N(E_F, 0) \simeq \frac{N}{E_F}$$

So,

$$N(E_F) = \frac{1}{2} N(E, 0) \simeq \frac{1}{2} \frac{N}{E_F}$$

Assuming that equal numbers of electrons have parallel and antiparallel spins, their number is given by

$$N(E_F)_s = \frac{1}{4} \frac{N}{E_F}$$

The approximation is made, for large n, that

$$\Delta E_N \simeq 2N(E_F)_s \, |\mu_B| \, H$$

From which

$$M \simeq -2N(E_F)_s \, \mu_B{}^2 H$$

Substituting the value for $N(E_F)_s$ gives

$$M \simeq -2 \frac{N}{4E_F} \mu_B{}^2 H = -\frac{N\mu_B{}^2 H}{2E_F} = -\frac{N\mu_B{}^2 H}{2k_B T_F}$$

and the diamagnetic susceptibility of the valence electrons as

$$\chi_D = \frac{M}{H} = -\frac{N\mu_B{}^2}{2k_B T_F} \tag{8-34}$$

Since E_F is virtually constant with temperature, χ_D behaves similarly. It is 1/3 that of the paramagnetic susceptibility of "free" electrons in normal metals (see Equation 8-126). χ_D Is of the order of 10^{-6} cm³/mol.

It is difficult to separate out this effect from the other effects present. These include the diamagnetic effects of the ion cores, Equations 8-32 or 8-33, and the paramagnetic effects of the valence electrons, (Equation 8-126) resulting from their spins.

The heavy metals, those having more electrons will have larger diamagnetic components than the light metals, and usually are diamagnetic for this reason. The net magnetic properties of gold, mercury, and lead are diamagnetic because of this. The paramagnetic effects predominate in many other metals.

The diamagnetic effect in metals is not expected to be a function of temperature.

However, at very low cryogenic temperatures, the susceptibilities of some metals oscillate as a function of field intensity. This variation is a periodic function of the reciprocal of the field strength. Such behavior is known as the de Haas-van Alpen effect. It has been shown that the periods of the oscillation give the minimum and maximum cross-sections of the Fermi surface normal to the field (see Figure 10-16, Chapter 10, Volume III).

8.3. PARAMAGNETISM

Paramagnetism results from some degree of alignment of the permanent ionic dipoles within a substance; this gives a positive susceptibility. As discussed in Section 8.2.1, many ionic and covalent compounds only show diamagnetic properties because their filled electron shells are electrically balanced and do not form dipoles. Magnetic dipoles only form when an electrical imbalance exists in the ions of a substance.

The simplest case of paramagnetic materials is that of the normal metals, those with filled inner shells and nearly free valence electrons in a quasi-continuum of energy states. The application of a magnetic field can affect relatively few of these electrons, those within about k_BT of E_F. This upsets the spin balance present in the absence of a field and a relatively weak, temperature-independent paramagnetism results (see Section 8.3.2).

Transition and rare-earth elements and their compounds show very pronounced paramagnetic behaviors. Ions of these elements have incomplete inner electron shells. The resulting electrical imbalance gives rise to strong, permanent magnetic moments. These constitute very effective dipoles in magnetic fields whose alignment is affected by thermally induced oscillations. Consequently, such substances show large, temperature-dependent paramagnetic behaviors (see Section 8.6).

Nuclei also have magnetic moments. However, the paramagnetic effects associated with nuclei are very small compared to the effects noted above.

As in the case of diamagnetism, the approach here will be to examine the classical analysis and then go on to the quantum-mechanical treatments of paramagnetism.

8.3.1. The Langevin Theory of Paramagnetism

The Langevin theory applies primarily to independent ions such as in substances composed of freely rotating dipoles, as gas molecules or liquids.

In the absence of an external field, a solid composed of freely moving noninteracting ions, or molecules, will consist of randomly oriented dipoles. The application of an external field to this material will exert a torque upon the dipoles; these will tend to align themselves parallel to the applied field. This represents a lower entropy condition since the molecules are ordered. The thermal vibrational activity of the ions or molecules comprising the material will tend to counteract this alignment in the field (Chapter 4, Volume I). This tendency toward disarray increases as the temperature increases because the oscillations increase and entropy increases as temperature rises. Eventually a temperature is reached at which the particles again become randomly oriented. The increasing randomness diminishes the paramagnetism which disappears when complete randomness occurs.

Consider a single dipole, such as that shown in Figure 8-4.

The potential energy, $E(\theta)$, of a dipole is defined such that it is zero when it is normal to the applied field, that is, for $\theta = 90°$. Where μ_M is the magnetic moment and H is the intensity of the applied field, the potential energy is given by

$$E(\theta) = -\vec{\mu}_M \cdot H = -\vec{\mu}_M H \cos\theta \tag{8-35}$$

The magnetic polarization within the substance is

$$M = N\mu_M \langle\cos\theta\rangle \tag{8-36}$$

in which N is the number of dipoles per unit volume and θ is the average dipole angle over a given distribution of dipoles.

The classic statistical probability of finding a dipole of potential energy $E(\theta)$ in an element of solid angle dV is determined from Equation 8-36, on the basis of the statistical average given by

$$\langle\cos\theta\rangle = \frac{\int_0^\pi \exp[-E(\theta)/k_BT]\cos\theta\, dV}{\int_0^\pi \exp[-E(\theta)/k_BT]\, dV} \tag{8-37}$$

The solid angle dV is the angle between θ and $\theta + d\theta$; it is measured by the surface area that it intercepts. A circle of unit radius, with its center at the origin, is used to help to reexpress the quantity dV. This makes use of

$$x^2 + y^2 = 1$$

$$x = \cos\theta$$

$$y = \sin\theta$$

$$dx = -\sin\theta\, d\theta$$

$$dy = \cos\theta\, d\theta$$

and the surface area is

$$s = 2\pi \int y \left\{1 + \left[\frac{dy}{dx}\right]^2\right\}^{1/2} dx$$

From these expressions, since the solid angle dV is measured by dS, and substituting for dy and dx,

$$dV = 2\pi \sin\theta \left[(dx)^2 + (dy)^2\right]^{1/2}$$

$$= 2\pi \sin\theta\, (\sin^2\theta + \cos^2\theta)^{1/2}\, d\theta$$

Since $\sin^2\theta + \cos^2\theta = 1$, this becomes

$$dV = 2\pi \sin\theta\, d\theta \tag{8-38}$$

The results given by Equation 8-38 are substituted into Equation 8-37 to obtain

$$\langle\cos\theta\rangle = \frac{\int_0^\pi \sin\theta\cos\theta \exp\left[-E(\theta)/k_BT\right] d\theta}{\int_0^\pi \sin\theta \exp\left[-E(\theta)/k_BT\right] d\theta} \tag{8-39}$$

where the factors 2π in the numerator and denominator cancel each other. Equation 8-35 is used in the exponentials to give

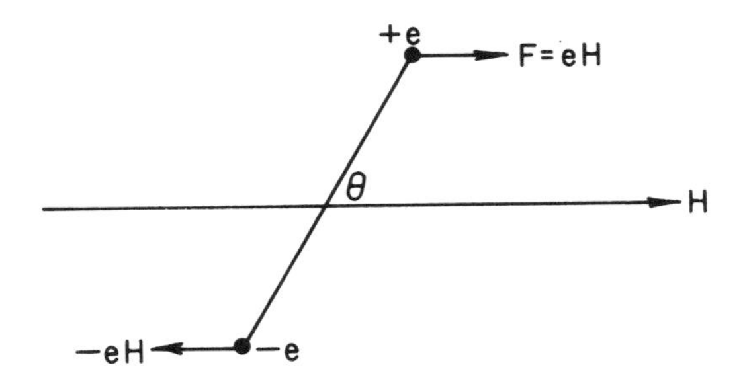

FIGURE 8-4. Sketch of a dipole.

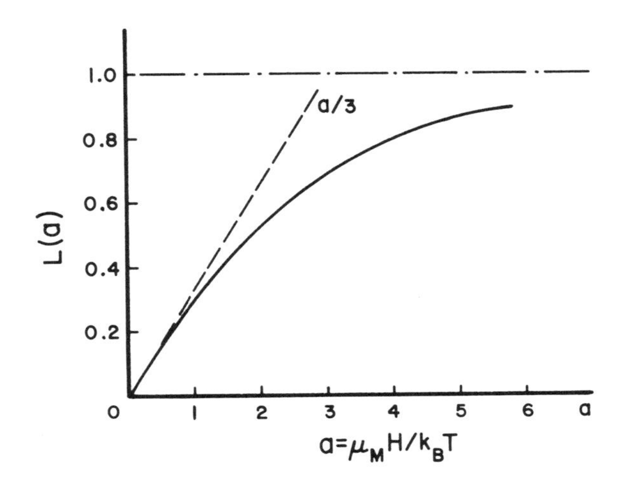

FIGURE 8-5. The Langevin function L(a) vs. a.

$$<\cos\theta> = \frac{\int_0^\pi \sin\theta\cos\theta\exp\left[\mu_M H\cos\theta/k_B T\right]d\theta}{\int_0^\pi \sin\theta\exp\left[\mu_M H\cos\theta/k_B T\right]d\theta} \tag{8-40}$$

This may be simplified by letting

$$\frac{\mu H}{k_B T}\cos\theta = a\cos\theta = x \tag{8-41}$$

and differentiating to obtain

$$d\theta = -\frac{dx}{a\sin\theta} \tag{8-42}$$

Equations 8-41 and 8-42 are then substituted into Equation 8-40. This gives

$$<\cos\theta> = \frac{\int_0^\pi \sin\theta\cos\theta\, e^x \frac{dx}{a\sin\theta}}{\int_0^\pi \sin\theta\, e^x \frac{dx}{a\sin\theta}} = \frac{\int_0^\pi \cos\theta\, e^x dx}{\int_0^\pi e^x dx} \tag{8-43}$$

Now, multiplying the numerator by a/a, and recalling Equation 8-41,

$$<\cos\theta> = \frac{\frac{1}{a}\int_0^{\pi} a\cos\theta\, e^x dx}{\int_0^{\pi} e^x dx} = \frac{1}{a}\frac{\int_{-a}^{a} xe^x\, dx}{\int_{-a}^{a} e^x\, dx} \tag{8-44}$$

From tables of integrals, it is found that

$$\int x \cdot e^{bx} = \frac{e^{bx}}{b^2}(bx - 1) = e^x(x - 1) \tag{8-45}$$

since, in the present case, $b = 1$. Applying Equation 8-45 to Equation 8-44 results in

$$<\cos\theta> = \frac{1}{a}\frac{e^a(a-1) - e^{-a}(-a-1)}{e^a - e^{-a}}$$

$$= \frac{\left(e^a + e^{-a}\right) - \frac{1}{a}\left(e^a - e^{-a}\right)}{e^a - e^{-a}}$$

Performing the division results in

$$<\cos\theta> = \frac{e^a + e^{-a}}{e^a - e^{-a}} - \frac{1}{a} \tag{8-46}$$

or,

$$L(a) = <\cos\theta> = \coth a - \frac{1}{a} \tag{8-47}$$

This is the Langevin function, L(a). A plot of this function vs. the parameter a is given in Figure 8-5.

It will be noted that for $a > 5$, L(a) becomes asymptotic. This corresponds to the dipoles approaching maximum alignment. For an approximate solution, assume $a = \mu_M H/K_B T \ll 1$. This means that $\mu_M H \ll k_B T$. In other words, if the field strength is small and the temperature is sufficiently high, this approximation can be made. It will be recalled that, for small x, an approximation can be obtained from the first two terms of the series

$$\coth x \simeq \frac{1}{x} + \frac{x}{3} +$$

So, Equation 8-47 becomes, by means of this approximation,

$$L(a) = \frac{1}{a} + \frac{a}{3} - \frac{1}{a} = \frac{a}{3} \tag{8-48}$$

and, from Equation 8-41,

$$L(a) = \frac{1}{3}\frac{\mu_M H}{k_B T} \tag{8-49}$$

The magnetic polarization is obtained from Equation 8-36, using $< \cos\theta> = L(a)$, and is

$$M = N\mu_M < \cos\theta > = N\mu_M \cdot \frac{1}{3}\frac{\mu_M H}{k_B T} = \frac{N\mu_M{}^2 H}{3k_B T} \tag{8-50}$$

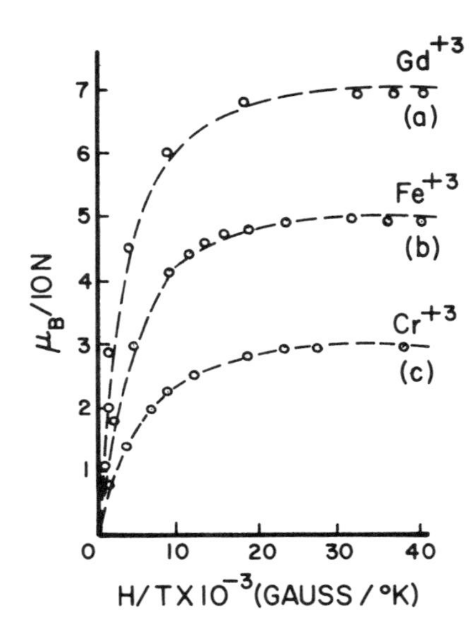

FIGURE 8-6. Magnetic moment per ion vs. H/T. (a) S = 7/2, (b) S = 5/2, (c) S = 3/2. See Section 8.3.2 for calculation of S and note that L = 0. Broken curves are L(a). (After Henry, W. E., *Phys. Rev.*, 88, 559, 1952. With permission.)

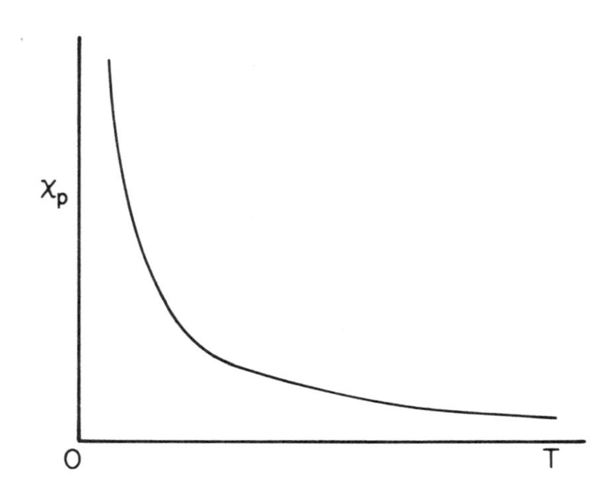

FIGURE 8-7. Paramagnetic susceptibility for the freely rotating dipole model used in the Langevin theory.

Since N is the number of dipoles per unit volume, the magnetic polarizability of a single dipole is $\mu_M{}^2/3k_BT$(see Equations 8-86, 8-87, and 8-88).

A plot of magnetization (Equation 8-50) as a function of H/T is given in Figure 8-6. Here, it can be seen that as the temperature increases, with H being held at some given value, the magnetization diminishes and approaches zero. This is in agreement with the earlier comment that the thermal activity of the dipoles tends to diminish their degree of alignment with the external field and, consequently, decreases the magnetization of the solid. In other words, as the entropy increases, the magnetic polarization of the substance diminishes. The maximum magnetic polarization, approaching 100%

alignment of the dipoles with the magnetic field, is attained at temperatures within a few degrees of 0 K.

Equation 8-50 permits the calculation of the paramagnetic susceptibility. Using the definition given by Equation 8-9 and the expression for the magnetization given by Equation 8-50 results in

$$\chi_p = \frac{M}{H} = \frac{1}{H}\cdot\frac{N\mu_M^{\,2}H}{3k_BT} = \frac{N\mu_M^{\,2}}{3k_BT} \qquad (8\text{-}51)$$

Unlike diamagnetism, which is independent of temperature (Equation 8-33) paramagnetic susceptibility is inversely proportional to the temperature. Like diamagnetic behavior, it is independent of the field (see Equation 8-85).

It will be noted that for a given material, $N\mu_M^2/3k_B$ is a constant which is known as the Curie constant, C. Thus, the Curie law may be written as

$$\chi_p = \frac{C}{T} \qquad (8\text{-}52)$$

$$C = N\mu_M^2/3k_B$$

The general behavior of χ_p is shown in Figure 8-7. The effect of temperature is more readily apparent for this property than for the magnetization shown in Figure 8-6.

8.3.2. Quantum Mechanic Treatment

The preceding treatment of paramagnetism was based upon the assumption that the dipoles were "free" to rotate in response to the applied field. In the cases where this is not possible, paramagnetic behavior results from the spin and angular momenta of the electrons. These factors can assume only quantized states in the solid. As such, they are not free in the sense used above. Thus, it is necessary to review and to extend some of the concepts developed in Section 3.11, Chapter 3, Volume I.

The principal quantum number, n, determines the "radius" and energy of a shell. The second quantum number is ℓ, and is an index of angular orbital momentum. This parameter varies from zero to (n−1) in integral steps. The angular orbital momentum is given in terms of ℓ by the vector, where $L = [\ell(\ell + 1)]^{1/2}$,

$$L_\ell = \frac{h}{2\pi}L, \quad \text{or} \quad L_\ell = \frac{h}{2\pi}[\ell(\ell + 1)]^{1/2} \qquad (8\text{-}53)$$

It will be recalled that ℓ determines the "shape" of an "orbit," so that it can be seen why this quantum number must be a measure of the angular orbital momentum. The angular magnetic moment of an electron is given when Equation 8-53 is substituted into Equation 8-18 to obtain

$$\mu_{M\ell} = \frac{e}{2mc}\cdot\frac{h}{2\pi}[\ell(\ell + 1)]^{1/2} \qquad (8\text{-}54)$$

This relationship also defines another fundamental parameter, the Bohr magneton, as

$$\mu_B = \frac{eh}{2mc} \qquad (8\text{-}55)$$

This has the values of 9.27×10^{-21} erg/Oersted, or 9.27×10^{-24} amp−(meter)^2.

The third quantum number is m_ℓ. It varies in integral steps from $-\ell$ to $+\ell$, including zero. It was noted previously that m_ℓ is related to ϕ. As such, it is a measure of the component of angular momentum about the z axis in Figure 3-10, Chapter 3, Volume I. This may be expressed as

$$p(\phi) = \frac{h}{2\pi} m_\ell ; \; m_\ell = 0, \pm 1, \pm 2, \ldots \pm \ell$$

so that m_ℓ, as discussed in Section 3.11, Chapter 3, Volume I, depends upon ℓ, and can have a multiplicity of $2\ell + 1$ values. The z axis is undefined in the absence of a magnetic field, and m_ℓ is degenerate. However, when a magnetic field is applied, the z axis is defined; the orbital quantum number, ℓ, may take on certain different orientations with respect to the z axis; m_ℓ no longer is degenerate. Each such orientation of ℓ must be such that its projection to the z axis results in quantized values of m_ℓ. In this way, m_ℓ defines the allowed orientations of the angular momentum with respect to the field. In other words, the presence of the field removes the degeneracy of m_ℓ. This is shown in Figure 8-8. The resulting values of m_ℓ give corresponding values in the electron energies where only one degenerate energy level was present prior to the application of the field. This accounts for the spectral line splitting, the Zeeman effect, previously noted in Section 3.11, Chapter 3, Volume I. (See the following section and Figure 8-13, Section 8.3.2.1.)

The remaining quantum number, m_s, or simply s, represents "spin." This also was postulated originally in order to explain spectra (Section 3.11 in Chapter 3, Volume I). It is not predicted by Equation 3-76, but the theoretical explanation for the necessity for this attribute was given later by Dirac (see Section 5.4.4, Chapter 5, Volume I). Spin can have only the values of $\pm 1/2$. This multiplicity can be expressed, in a way corresponding to that for ℓ, as $2s + 1$. The angular momentum of an electron resulting from its intrinsic spin is given by the vector, where $S = [s(s + 1)]^{1/2}$,

$$L_s = \frac{h}{2\pi} S$$

$$\text{or} \quad L_s = \frac{h}{2\pi} [s(s + 1)]^{1/2} \tag{8-56}$$

This is shown in Figure 8-9. Again using Equation 8-18, the component of the magnetic moment due to spin is given by the substitution of Equation 8-56 to obtain

$$\mu_{Ms} = \frac{e}{2mc} \cdot \frac{h}{2\pi} [s(s + 1)]^{1/2} = \mu_B [s(s + 1)]^{1/2} \tag{8-57}$$

Thus, for example, for $s = 1/2$, $\mu_{Ms} = (\sqrt{3/2})\mu_B$

The total momentum of an electron is the sum of its angular and intrinsic momenta. These two components are taken into account by the vector sum, J, known as the total, or inner, quantum number. The total quantum number is given by $J = [j(j + 1)]^{1/2}$ and the total momentum by

$$L_j = \frac{h}{2\pi} [j(j + 1)]^{1/2} \tag{8-58}$$

in which $j = \ell \pm s$. Two examples of calculating j are given in Figure 8-10, in the absence of a magnetic field.

In a magnetic field L, S, and J are oriented relative to the field and precess about the direction of the field. This dynamic behavior is known as Russell-Saunders coupling. This is shown in Figure 8-11 for a single electron.

More complex behavior than that shown in the figure can occur when strong inter-

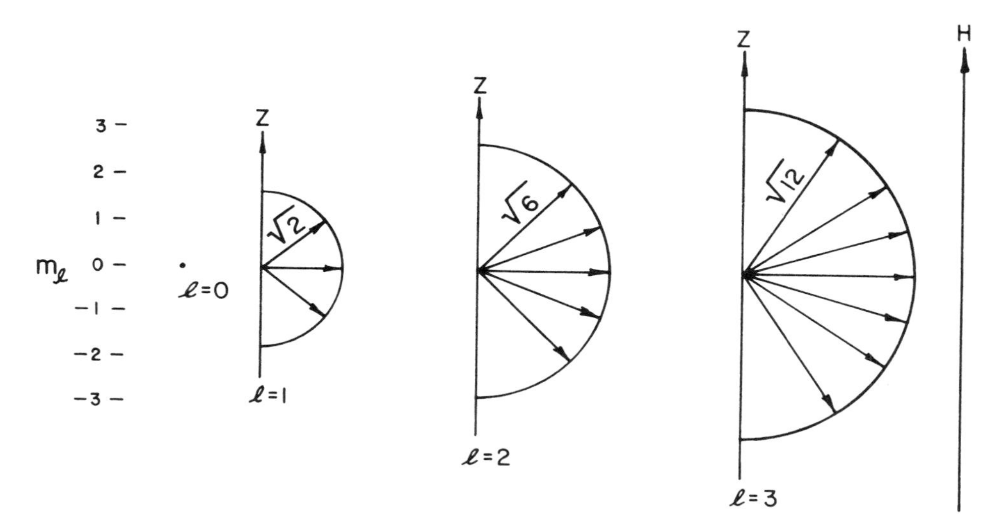

FIGURE 8-8. Quantized states of m_ℓ which result from the permissible orientations of the angular momentum in relation to an applied magnetic field.

actions take place between the spin and orbital vectors of a number of electrons. This can take place in the d bands of transition elements. The $s_i - s_j$ and $\ell_i - \ell_j$ couplings may be large compared to $s_i - \ell_j$ interactions among electrons. When this is the case, $S = \Sigma s_i$ and $L = \Sigma \ell_i$ and the sum of these combined vectors gives the diagram for the resultant, J; this is more complicated than that shown in Figure 8-11.

In either the simple or interactive cases discussed above, the total quantum number is given by

$$J = L + S \tag{8-59}$$

The rules governing the sums of quantum numbers given by Equation 8-59 are known as Hund's rules. The first of these states that the value of S is maximized by adding spins insofar as permitted by the exclusion principle. This means that the electrons are in different states. Second, the orbital momenta are maximized to give the largest L which is in agreement with the first rule. The implication here is that the electrons are orbiting in the same direction. These lead to a third rule which explicitly defines Equation 8-59, namely;

$$J = L + S \tag{8-59a}$$

and

$$J = L - S \tag{8-59b}$$

Equation 8-59a applies to bands which are more than half-filled because of the parallel coupling of S and L; Equation 8-59b applies to bands which are less than half-filled because of the antiparallel coupling.

The following examples of these rules are given for 3d electrons: $S = 3/2$ and $5/2$ for 3 and 5 such electrons, respectively. Where the band is more than half-filled with 6 electrons, $S = 5/2 - 1/2 = 2$ because of the antiparallel spin in the second half of the band. A completely filled d band would have $S = 0$. To illustrate the second rule, three 3d electrons would have $L = 2 + 1 + 0 = 3$ for maximum L. For five electrons,

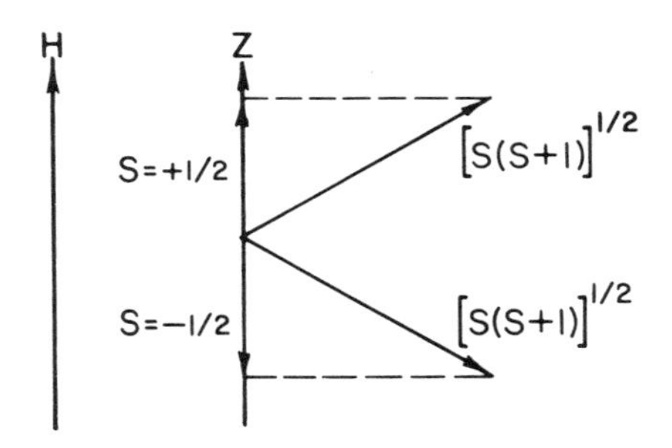

FIGURE 8-9. Projected values of spin parallel to an applied magnetic field.

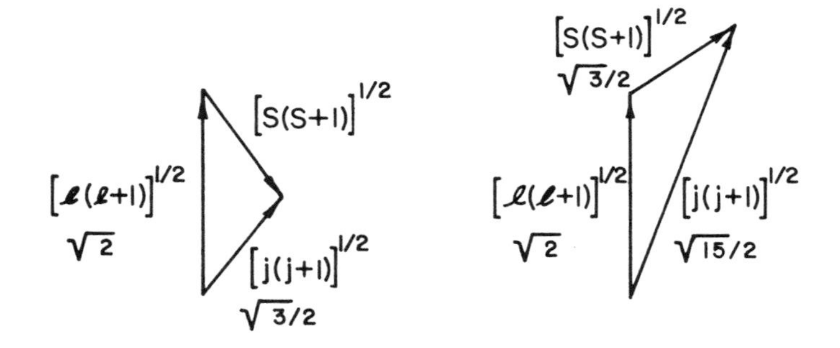

FIGURE 8-10. Examples of electron momenta vectors for $\ell = 1$ and s = 1/2 for a single electron in a hydrogen-like atom (units of $h/2\pi$).

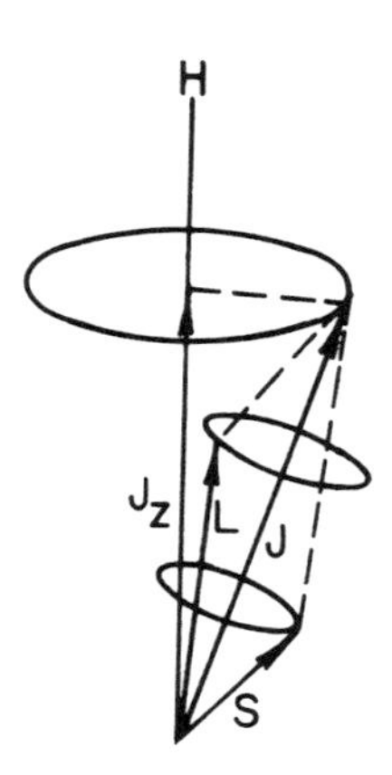

FIGURE 8-11. Diagram of Russell-Saunders coupling.

L = 2 + 1 + 0 − 1 − 2 = 0, so that a half-filled d band would have a zero value for L. For six electrons, L = 2 + 1 + 0 − 1 − 2 + 2 = 2. For three d electrons, J = 2 + 1 + 0 − 3/2 = 3/2; for six d electrons, J = 2 + 1 + 0 − 1 − 2 + 2 + (5/2 − d½) = 4.

On the basis of angular momentum alone, the magnetic moment of an electron is given by Equation 8-54. This was the basis for obtaining the expression for the Bohr magneton. This is not necessarily the magnetic moment of an electron. As has been shown, the magnetic moment of an electron must take both orbital and spin components into account. In other words, J must be considered. This is developed in the next section.

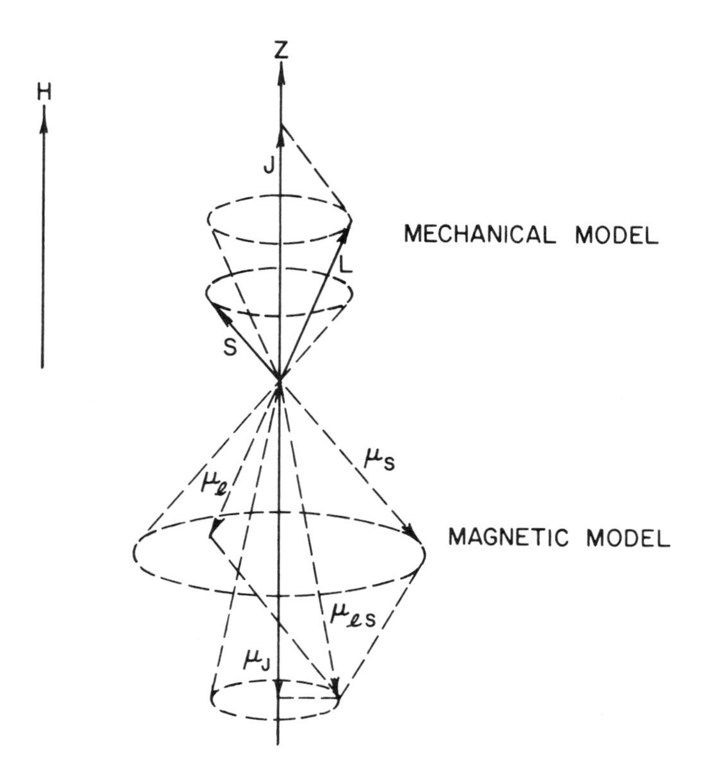

FIGURE 8-12. Diagram of mechanical and magnetic models for the Landé spectrographic splitting factor. (After White, H. E., *Introduction to Atomic Spectra*, McGraw-Hill, New York, 1934, 155. With permission.)

8.3.2.1. Landé Spectroscopic Splitting Factor

The Zeeman effect (Figure 8-13) is observed in the presence of magnetic fields. In other words, when the energy of the applied field is considerably smaller than that associated with an electron, the spectral line, representing an energy level, splits into two or more lines, or levels. This results from the L-S coupling. The Landé factor measures the interaction and determines the way a level splits. A diagram of the mechanical and magnetic models involved is shown in Figure 8-12.

In the magnetic model, $\mu_{\ell S}$ is not colinear with the mechanical resultant, J. In the diagram, $\mu_\ell = -\mu_{B}\ell$ for orbital motion. But an electron with s = ½ has one Bohr magneton, not one-half of a Bohr magneton. Thus, $\mu_s = 2\mu_B s$. Therefore, the resultant, $\mu_{\ell_s} = \mu_\ell + \mu_s = -\mu_B(\ell + 2s)$ is not colinear with J. The component of $\mu\ell_s$ which is parallel to the field, μ_j, is colinear with J.

The components S, L, μ_s, μ_{ℓ_s} precess about J. Because of this, only the component of μ_{ℓ_s} which is parallel to J contributes to the magnetic moment. The component perpendicular to J averages to zero because of the precession. The magnetic moments are

$$\mu_S = 2S\,\frac{e\hbar}{2mc} \tag{8-60}$$

and

$$\mu_\ell = L\,\frac{e\hbar}{2mc} \tag{8-61}$$

Their components parallel to J are, using the notation (U; V) as the cosine of the angle between the vectors U and V *(Note that this is not the vector dot product.)*

$$\mu_{sj} = 2S \frac{e\hbar}{2mc} (S;J) \tag{8-62}$$

and

$$\mu_{\ell j} = L \frac{e\hbar}{2mc} (L;J) \tag{8-63}$$

The vector sum of these components is

$$\mu_{sj} + \mu_{\ell j} = \mu_j = [2S(S;J) + L(L;J)] \frac{e\hbar}{2mc} \tag{8-64}$$

The quantity within the brackets is evaluated by equating it to Jg, where g is the spectroscopic splitting factor. Thus,

$$Jg = 2S\,(S;J) + L(L;J) \tag{8-65}$$

This can be simplified by recalling the law of cosines, namely:

$$a^2 = b^2 + c^2 - 2bc \cos A$$

From this

$$S^2 = L^2 + J^2 - 2LJ\,(L;J)$$

and

$$L(L;J) = \frac{J^2 + L^2 - S^2}{2J} \tag{8-66}$$

Similarly,

$$S(S;J) = \frac{S^2 + J^2 - L^2}{2J} \tag{8-67}$$

Equations 8-66 and 8-67 are substituted into Equation 8-65 to obtain

$$Jg = 2\left[\frac{S^2 + J^2 - L^2}{2J}\right] + \frac{J^2 + L^2 - S^2}{2J} \tag{8-68}$$

and simplifying,

$$Jg = \frac{S^2 + 3J^2 - L^2}{2J} \tag{8-69}$$

and rearranging to get

$$g = \frac{S^2 + 3J^2 - L^2}{2J^2} = \frac{2J^2 + J^2 + S^2 - L^2}{2J^2}$$

and finally by dividing

$$g = 1 + \frac{J^2 + S^2 - L^2}{2J^2} \tag{8-70}$$

Additional information is required for the solution of Equation 8-70. The relationships in the previous section are used for this. These are given in the squared form as $L^2 = L(L+1)$, $S^2 = S(S+1)$ and $J^2 = J(J+1)$ where $J = L + S$ and L, S, and J are calculated as shown at the end of the previous section. When these substitutions are made, the Landé spectrographic splitting factor is given by

$$g = 1 + \frac{J(J+1) + S(S+1) - L(L+1)}{2J(J+1)} \tag{8-71}$$

It will be seen that $g = 2$ for electron spin alone, when $L = 0$. It equals 1 when only L is involved and when spin is not, that is, when $S = 0$.

J precesses in a magnetic field as shown in Figure 8-11. Its component parallel to the field, J_z, is quantized just as m_ℓ is. J_z can have $2J + 1$ values, an important factor in both paramagnetism and ferromagnetism.

The ratio of Equation 8-64 to Equation 8-65 is taken to give

$$\frac{\mu_j}{Jg} = \frac{[2S(S;J) + L(L;J)]\dfrac{e\hbar}{2mc}}{2S(S;J) + L(L;J)} \tag{8-72}$$

which simplifies to give the ratio of the magnetic and mechanical moments as

$$\frac{\mu_j}{J} = g\frac{e\hbar}{2mc} = g\mu_B \tag{8-73}$$

In a magnetic field J precesses around the z axis and gives a quantized value for J_z parallel to the field as shown in Figure 8-11. The corresponding value of J_z is shown in this figure, but both J and J_z are shown as being colinear in Figure 8-12. Thus, Equation 8-73 may be expressed in more general terms as

$$\mu_z = g\mu_B J_z \tag{8-74}$$

The spectrographic line splitting is given as an example for the case in which only spin momenta are involved. This is based upon the behavior noted in Figure 8-9 and is shown in Figure 8-13a, along with the case for $J = 3/2$. Using Equation 8-74, the difference in the energies of the split levels is given by

$$\Delta E = 2|\mu_z|H; \quad \left\{\frac{\text{ergs}}{\text{Oersted}} \times \text{Oersted} = \text{ergs}\right\} \tag{8-75}$$

The factor 2 is present because J_z can have both positive and negative values. The absolute values of μ_z are used because of this; the line splitting occurs in two directions. The equation gives

$$\Delta E = 2|g\mu_B J_z|H = 2\left|2\mu_B\frac{1}{2}\right|H = 2\mu_B H \tag{8-75a}$$

since, for spin alone, $g = 2$ and $J_z = ½$ in Figure 8-13a, giving agreement with the observed behavior for this case. The last two equations also may be used to verify the value of g for spin. Equating these gives

$$2|\mu_z|H = 2|g\mu_B J_z|H$$

and

$$|\mu_z| = |g\mu_B J_z|$$

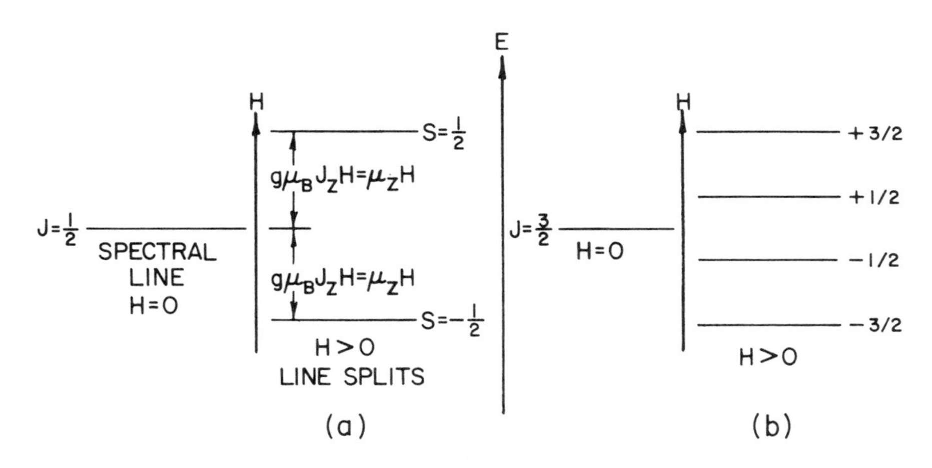

FIGURE 8-13. The Zeeman effect for an electron. (a) When spin alone is considered; (b) when spin and angular orbital momenta are involved. Only 2J + 1 states are permitted.

For electron spin $\mu_z = \mu_B$ and $J_z = 1/2$, which gives g = 2.

The expression for the Larmor precession (Equation 8-26) only took orbital motion into consideration. Thus, in this case, S = 0 and g = 1, so

$$\omega_L = -\frac{eH}{2mc} = \gamma H; \; \gamma = -\frac{e}{2mc} = -\frac{\mu_B}{\hbar} \quad (8\text{-}76a)$$

Here γ is the gyromagnetic ratio, and

$$\omega_L = -\frac{\mu_B}{\hbar} H \quad (8\text{-}76b)$$

When a spin component adds to the orbital momentum, the gyromagnetic ratio becomes

$$\gamma = -\frac{ge}{2mc} = -\frac{g\mu_B}{\hbar} \quad (8\text{-}77a)$$

then, the Larmor precession is given by

$$\omega_L = -\frac{g\mu_B}{\hbar} H \quad (8\text{-}77b)$$

The application of an oscillating field of frequency given by Equation 8-77b can induce electron transitions to different levels because of the absorption of energy from the field. This reaction is known as paramagnetic, or electron spin, resonance because it occurs at, or close, to ω_L.

8.3.3. Paramagnetic Ions in Compounds

When a substance is composed of ions containing unpaired electrons, it will show paramagnetic behavior. In the presence of a magnetic field these electrons will split into two groups in a way similar to that shown in Figure 8-13. The fractions with positive and negative moments, at thermal equilibrium, are given, where N_+ is the number with positive moments, N_- the number with negative moments and N is the total number of ions, by

$$\frac{N_+}{N} = \frac{\exp[\mu H/k_B T]}{\exp[\mu H/k_B T] + \exp[-\mu H/k_B T]} \quad (8\text{-}78a)$$

and

$$\frac{N_-}{N} = \frac{\exp[-\mu H/k_BT]}{\exp[\mu H/k_BT] + \exp[-\mu H/k_BT]} \quad (8\text{-}78b)$$

Equation 8-74 is used to substitute for μ in the exponentials to give

$$N_+ = N\frac{\exp[gJ\mu_BH/k_BT]}{\exp[gJ\mu_BH/k_BT] + \exp[-gJ\mu_BH/k_BT]}$$

$$= N\frac{e^x}{e^x + e^{-x}} \quad (8\text{-}79a)$$

where $x = gJ\mu_BH/k_BT$, and in a similar manner,

$$N_- = N\frac{e^{-x}}{e^x + e^{-x}} \quad (8\text{-}79b)$$

The fractions N_+/N and N_-/N are shown in Figure 8-14.

The imbalance between the numbers of positive and negative moments is just the difference between N_+ and N_-, or

$$N_+ - N_- = N\left[\frac{e^x}{e^x + e^{-x}} - \frac{e^{-x}}{e^x + e^{-x}}\right] = N\frac{e^x - e^{-x}}{e^x + e^{-x}} \quad (8\text{-}80)$$

This may be reexpressed as

$$N_+ - N_- = N \tanh x \quad (8\text{-}81)$$

Each electron has a magnetic moment of $gJ|\mu_B|$, Equation 8-74. The magnetic moment in the substance is determined by the imbalance between the two groups with opposite moments. The magnetization is given by

$$M = (N_+ - N_-)\mu_z = NgJ|\mu_B| \tanh x \quad (8\text{-}82)$$

For the condition similar to that used to approximate the Langevin function, Equation 8-47, i.e., for weak fields and sufficiently high temperatures,

$$\mu_BH \ll k_BT$$

then

$$x = \frac{gJ\mu_BH}{k_BT} \ll 1 \quad (8\text{-}83)$$

Under these conditions $\tanh x \simeq x$, and Equation 8-82 becomes

$$M = NgJ|\mu_B|x = NgJ|\mu_B| \cdot \frac{gJ\mu_BH}{k_BT} = \frac{Ng^2J^2\mu_B^2H}{k_BT} \quad (8\text{-}84)$$

From this the paramagnetic susceptibility is found to be, setting $g = 2$ and $J = \frac{1}{2}$,

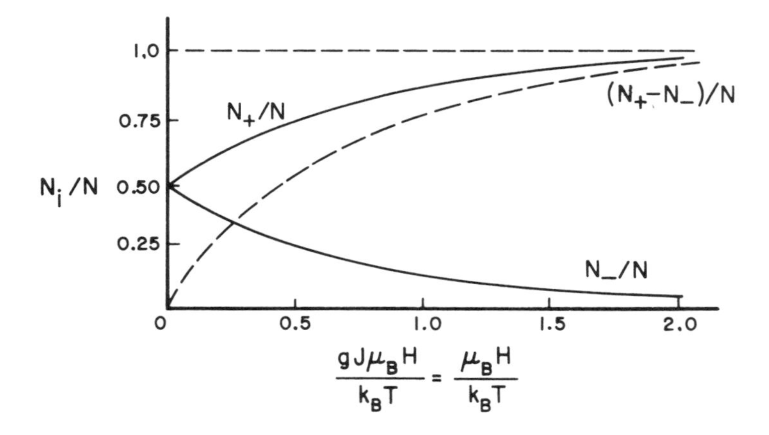

FIGURE 8-14. Fractions of ions with positive and negative spin moments for g = 2 and J = 1/2. Ions of fraction N_+/N have moments in the same direction as the field; the fraction N_-/N has moments in the direction opposite to the field.

$$\chi_P = \frac{M}{H} = \frac{Ng^2J^2\mu_B^2}{k_BT} = \frac{N\mu_B^2}{k_BT} \tag{8-85}$$

This is the quantum mechanic result for a substance with two spin orientations.

It will be recalled from Equation 8-51 that

$$\chi_P = \frac{N\mu^2{}_M}{3k_BT} \tag{8-51}$$

was given by the classical, Langevin, theory. It is of interest to compare these results with those given by the quantum approach:

$$\chi_P = \frac{N\mu^2{}_M}{3k_BT} = \frac{Ng^2J^2\mu_B^2}{k_BT} = \frac{N\mu_B^2}{k_BT} \tag{8-86}$$

From this it is seen that

$$\mu_B^2 = \frac{\mu^2{}_M}{3} \tag{8-87}$$

This gives an apparent correspondence between the classical and quantum mechanic findings.

If, however, the effective number of Bohr magnetons is given by n_B, the effective magnetic moment is expressed as

$$\mu^2{}_M = n^2{}_B\mu_B^2 \tag{8-88}$$

Equation 8-104a defines $n^2{}_B = g^2[J(J+1)] = 4 \times 3/4 = 3$ for spin. When this is taken into account, the Langevin equation becomes

$$\chi_P = \frac{Nn^2{}_B\mu_B^2}{3k_BT} = \frac{3N\mu_B^2}{3k_BT} = \frac{N\mu_B^2}{k_BT} \tag{8-89}$$

Thus, the Langevin and quantum expressions are identical, obeying the Bohr correspondence principle, and are of the form of the Curie law:

$$\chi_P = \frac{C}{T} \tag{8-90}$$

where $C = Nn^2_B\mu^2_B/3k_B$ or $N\mu^2_B/k_B$. Equation 8-85 is the quantum mechanic equivalent of the Langevin equation. The behavior of the paramagnetic susceptibility of a substance of this kind is shown in Figure 8-7.

8.3.4. χ_P in Terms of J

This derivation of paramagnetic susceptibility is based upon the total, or inner, quantum number, J, as shown in Figure 8-11 and actually is J_z. It will be recalled that this quantity is given by the sum of L and S (Equation 8-59, Section 8.3.2). As such, it combines orbital and intrinsic angular moments and gives more insight into the mechanism.

The magnetic moment is, from Equation 8-74,

$$\mu_j = Jg\mu_B \tag{8-91}$$

The potential energy is

$$\text{P.E.} = -\mu_j H \tag{8-92}$$

The average magnetization for N atoms per unit volume, using the Boltzmann statistics, is obtained from

$$M = N \frac{\sum_{-J}^{J} Jg\mu_B \exp\left[Jg\mu_B H/k_B T\right]}{\sum_{-J}^{J} \exp\left[Jg\mu_B H/k_B T\right]} \tag{8-93}$$

Solutions to Equation 8-93 will be considered for two cases.

The first solution to Equation 8-93 is based upon

$$Jg\mu_B H/k_B T < 1 \tag{8-94}$$

Using the series approximation for the exponential term, Equation 8-93 becomes

$$M = Ng\mu_B \frac{\sum_{-J}^{J} J\left[1 + \frac{Jg\mu_B H}{k_B T}\right]}{\sum_{-J}^{J}\left[1 + \frac{Jg\mu_B H}{k_B T}\right]} \tag{8-95}$$

Consider the numerator, letting $g\mu_B H/k_B T = x$. This becomes

$$\sum_{-J}^{J} J(1 + Jx) \tag{8-96}$$

The total quantum number J can assume values of

$$J,\ (J-1),\ (J-2),\ \ldots,\ -(J-2),\ -(J-1)\ -J$$

or a total of 2J + 1 values (Section 8.3.2.1.). Now, Equation 8-96, upon substitution, is given by

$$\sum_{-J}^{J} J(1+Jx) = J(1+Jx) + (J-1)\left[1+(J-1)x\right] + \ldots - (J-1)\left[1-(J-1)x\right] - J\left[1-(-J)x\right]$$

or

$$\sum_{-J}^{J} J(1+J) = J + J^2x + (J-1) + (J-1)^2x \ldots - (J-1) + (J-1)^2x - J - J^2x$$

The terms involving the first power of J cancel out leaving only $(J-n_i)^2$ terms:

$$\sum_{-J}^{J} J(1+Jx) = J^2x + (J-1)^2x + \ldots + \left[-(J-1)^2\right]x + \left[-J\right]^2x$$

The Jn_i and n_i terms vanish and the numerator of Equation 8-95 is

$$\sum_{-J}^{J} J(1+Jx) = 2\sum_{-J}^{J} J^2x \qquad (8\text{-}97)$$

Now consider the denominator of Equation 8-95 expanded as a series:

$$\sum_{-J}^{J} (1+Jx) = 1+Jx + 1 + (J-1)x + \ldots + 1 - (J-1)x + 1 + (-J)x$$

Here all of the terms in J cancel out. And, since there are 2J + 1 terms

$$\sum_{-J}^{J} (1+Jx) = \sum_{-J}^{J} (1) = 2J + 1 \qquad (8\text{-}98)$$

When Equations 8-97 and 8-98 are substituted back into Equation 8-95, it becomes

$$M = Ng\mu_B \frac{2\sum_{-J}^{J} J^2x}{2J+1} = Ng^2\mu_B^2 \frac{H}{k_BT} \frac{2\sum_{-J}^{J} J^2}{2J+1} \qquad (8\text{-}99)$$

The numerator can be reexpressed by the sum of the series as

$$2\sum_{-J}^{J} J^2 = 2\cdot\frac{J}{6}(J+1)(2J+1)$$

Thus, Equation 8-99 becomes

$$M = Ng^2\mu_B^2 \frac{H}{k_BT}\cdot\frac{J(J+1)(2J+1)}{3(2J+1)} \qquad (8\text{-}100)$$

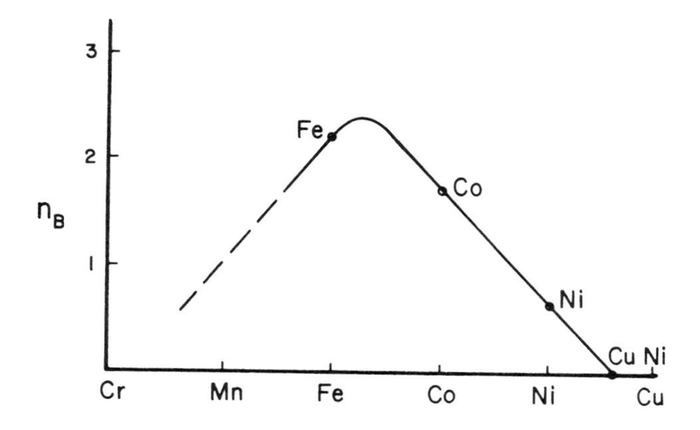

FIGURE 8-15. Effective number of Bohr magnetons of the first series of transition elements.

Upon simplification,

$$M = Ng^2\mu_B^2 \frac{H}{k_BT} \cdot \frac{J(J+1)}{3} = \frac{Ng^2J(J+1)}{3k_BT}\mu_B^2H \qquad (8\text{-}101)$$

$$= \frac{N\mu_B^2H}{k_BT}$$

for spin alone. Then, the paramagnetic susceptibility is found to be

$$\chi_P = \frac{M}{H} = \frac{Ng^2J(J+1)\mu_B^2}{3k_BT} = \frac{C}{T} = \frac{N\mu_B^2}{k_BT} \qquad (8\text{-}102)$$

for spin alone. This is another expression for the Curie law. It reduces to Equation 8-85 for spin ($g = 2$, $J = ½$). The effective magnetic moment is obtained from Equation 8-102, in the squared form, as

$$\mu^2_j = g^2J(J+1)\mu_B^2 \qquad (8\text{-}103)$$

The effective number of Bohr magnetons, n_B, may be obtained from Equation 8-103 starting with

$$n^2_B = g^2[J(J+1)] \qquad (8\text{-}104a)$$

which results in

$$n_B = g[J(J+1)]^{1/2} \qquad (8\text{-}104b)$$

This value for the effective number of Bohr magnetons must also be used in Equations 8-88 and 8-89, where it was employed, but not derived.

Figure 8-15 gives the value of n_B in terms of S rather than J. Here, $n_B = g[S(S+1)]^{1/2}$. These ions of transition elements behave as though no electron orbital effects are operative, and that spin alone is present. Since the 3d levels are outermost, they appear to react with the equivalent of an internal field and are not affected by an external field. This is known as quenched orbital momentum caused by a crystal field. In such cases, spin alone is responsible for the observed behavior (see Section 8.6.2 and compare it with data in Section 8.6.1).

The nonintegral values of n_B of the elements in Figure 8-15 result from the s-d transitions discussed in Sections 7.8.3 and 7.8.4, Chapter 7. The effects of the alloying elements upon the degree of electron occupation of the d bands (Sections 7.11.1 and 7.11.2) are responsible for the resulting changes in spin imbalance; this accounts for the variations in n_B of the alloys between the elements shown.

The second solution in Equation 8-93 is for strong applied fields and low temperatures. Equation 8-93 is written in the form

$$M = Ng\mu_B \frac{\sum_{-J}^{J} Je^{Jx}}{\sum_{-J}^{J} e^{Jx}} \tag{8-105}$$

where, as previously, $x = g\mu_B H/k_B T$.

It will be noted that Equation 8-105 can be expressed as

$$M = Ng\mu_B \frac{d}{dx}\left[\ln \sum_{-J}^{J} e^{Jx}\right] \tag{8-106}$$

Expanded in individual terms, the summation is written as

$$\begin{aligned}\sum_{-J}^{J} e^{Jx} &= e^{Jx} + e^{(J-1)x} + {}^{(J-2)x} +++ e^{-(J-2)x} + e^{-(J-1)x} + e^{-Jx} \\ &= e^{Jx} + e^{Jx-x} + e^{Jx-2x} +++ e^{-Jx+2x} + e^{-Jx+x} + e^{-Jx} \\ &= e^{Jx}\left[1 + e^{-x} + e^{-2x} +++ e^{-2Jx+2x} + e^{-2Jx+x} + e^{-2Jx}\right]\end{aligned} \tag{8-107}$$

Substituting Equation 8-107 back into Equation 8-106 gives

$$M = Ng\mu_B \frac{d}{dx}\left\{\ln\left[e^{Jx}\left(1+e^{-x} + e^{-2x} +++ e^{-2Jx+2x} + e^{-2Jx+x} + e^{-2Jx}\right)\right]\right\} \tag{8-108}$$

Summing the geometric series results in a simpler expression:

$$M = Ng\mu_B \frac{d}{dx}\left[\ln e^{Jx} \cdot \frac{1 - e^{-(2J+1)x}}{1 - e^{-x}}\right] = Ng\mu_B \frac{d}{dx}\left[\ln f(J,x)\right] \tag{8-109}$$

The quantity within the brackets may be rewritten as

$$\ln f(J,x) = \ln \frac{e^{Jx} - e^{-(J+1)x}}{1 - e^{-x}}$$

Multiplying the numerator and denominator of this fraction by $e^{x/2}$ gives

$$\ln f(J,x) = \ln \frac{e^{(J+1/2)x} - e^{-(J+1/2)x}}{e^{x/2} - e^{-x/2}}$$

Recalling that $\sinh u = e^{u} - e^{-u})/2$, the logarithm can be reexpressed as

$$\ell n f(J,x) = \ell n \frac{\sinh[(J+1/2)x]}{\sinh(x/2)} = \ell n \sinh[(J+1/2)x] - \ell n \sinh(x/2)$$

The derivative is found to be

$$\frac{d}{dx}[\ell n f(J,x)] = \frac{(J+1/2)\cosh[(J+1/2)x]}{\sinh[(J+1/2)x]} - \frac{1/2 \cosh(x/2)}{\sinh(x/2)}$$

or

$$\frac{d}{dx}[\ell n f(J,x)] = (J+1/2)\coth[(J+1/2)x] - 1/2 \coth(x/2)$$

Now, let a = Jx and, substituting for x, the derivative becomes

$$\frac{d}{dx}\left[\ell n f(J,x)\right] = \frac{2J+1}{2}\coth\left[\frac{2J+1}{2J}a\right] - \frac{1}{2}\coth\left[\frac{a}{2J}\right] \quad (8\text{-}110)$$

When multiplied by J/J, Equation 8-110 becomes

$$\frac{d}{dx}\left[\ell n f(J,x)\right] = J\left\{\frac{2J+1}{2J}\coth\left[\frac{2J+1}{2J}a\right] - \frac{1}{2J}\coth\left[\frac{a}{2J}\right]\right\} = JB_J(a) \quad (8\text{-}111)$$

The quantity within the brackets is the Brillouin function, $B_J(a)$. By means of Equation 8-111, Equation 8-106 becomes

$$M = Ng\mu_B JB_J(a) = Ng\mu_B B_J(Jg\mu_B H/k_B T) \quad (8\text{-}112)$$

recalling that $a = Jx = Jg\mu_B H/k_B T$ (see Figures 9-1 and 9-3, in Chapter 9).

In the case where J is very large, that is, where all orientations of J are allowed in the applied field, $B_J(a)$ reduces to the Langevin function, Equation 8-47, in agreement with the Bohr correspondence principle. For large J, the first term in Equation 8-111 reduces to coth a. The second term may be approximated by using the first term of the series

$$\coth u = \frac{1}{u} + \frac{u}{3} - \frac{u^3}{45} \ldots \simeq \frac{1}{u}$$

because the remaining terms are negligibly small in these circumstances. Thus, the second term becomes

$$1/2J \coth(a/2J) \simeq 1/2J \cdot 2J/a = 1/a$$

Thus, for large J,

$$B_J(a) \simeq \coth a - 1/a = L(a)$$

In strong fields at low temperatures, the second term in this approximation for $B_J(a)$ is negligible, and $B_J(a) \simeq \coth a \simeq a$. Under these conditions, Equation 8-112 becomes

$$M = NgJ\mu_B B_J(a) \simeq NgJ\mu_B \cdot Jg\mu_B H/k_B T = \frac{Ng^2 J^2 \mu_B{}^2 H}{k_B T} \quad (8\text{-}112a)$$

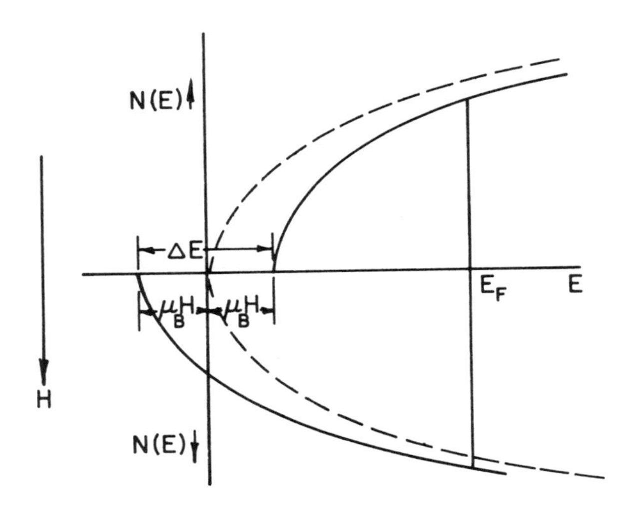

FIGURE 8-16. Effect of an applied magnetic field upon the density of states of a normal metal.

This is the same as Equation 8-84. And, where spin alone is concerned, becomes

$$\chi_p = \frac{M}{H} = \frac{N\mu^2 B}{k_B T} \tag{8-113}$$

Thus, the same results as those of Equations 8-85 and 8-89 are obtained.

8.3.5. Paramagnetic Susceptibility of Normal Metals

The case of normal metals is different from those previously discussed. Here, the nearly free valence electrons are responsible for the paramagnetic behavior; this is virtually independent of temperature. This is in contrast to the behavior where $\chi \propto 1/T$, as in Equations 8-51, 8-85, and 8-89. The difference in behavior was explained by Pauli and is known as Pauli paramagnetism.

When a magnetic field is applied to a normal metal, only a few electrons with energies of the order of $k_B T$ about E_F can be affected by the field. To a first approximation, the fraction of electrons which can be excited by the field is T/T_F (see Sections 5.3 and 5.5, Chapter 5, Volume I). When this is applied to Equation 8-51, the paramagnetic susceptibility is approximated by

$$\chi_p \simeq \frac{N\mu_B^2}{3k_B T} \cdot \frac{T}{T_F} = \frac{N\mu_B^2}{3k_B T_F} \tag{8-114}$$

It will be noted that Equation 8-114 is virtually a constant for a given material; it is independent of temperature to an extent corresponding to that of E_F and it is of the correct order of magnitude.

The electrons can be separated into two groups: those with spin parallel and antiparallel to an applied field. Their behavior in the absence of a field is shown by the dotted lines in Figure 8-16. However, when a field is applied, the Zeeman effect takes place (Figure 8-13). The single band splits into two half bands. These are shown by the solid lines in the fiture. One half band flips to lower energies and the other to higher energies than they occupied prior to the application of the field. Since the numbers of the affected electrons are proportional to the areas under the respective curves, the net imbalance between the two spins is apparent. The net excess of parallel spins over antiparallel spins is responsible for the paramagnetic behavior.

The number of electrons in the half band with spins parallel to the applied field, for $T \ll T_F$, is approximated by

$$N_+ = 1/2 \int_{-\mu_B H}^{E_F} f(E) N(E+\mu_B H) dE$$

$$\simeq 1/2 \int_{o}^{E_F} f(E) N(E) dE + 1/2 \mu_B H N(E_F) \qquad (8\text{-}115)$$

Similarly, the number of electrons in the half band with opposite spin is taken as

$$N_- = 1/2 \int_{\mu_B H}^{E_F} f(E) N(E-\mu_B H) dE$$

$$\simeq 1/2 \int_{o}^{E_F} f(E) N(E) dE - 1/2 \mu_B H N(E_F) \qquad (8\text{-}116)$$

Both of these approximations (Equations 8-115 and 8-116) give the lower limits as zero because $\mu_B H \ll E_F$ and the number of states between zero and $|\mu_B|H$ is small. Both also consider E_F to be a constant (Equation 5-26).

The number of electrons which flip from one band to the other is given by the difference between Equations 8-115 and 8-116 as

$$\Delta N = N_+ - N_- = \mu_B H N(E_F) \qquad (8\text{-}117)$$

One half band loses ΔN electrons, the other gains a like number. Thus, $2\Delta N$ is the difference in the numbers of electrons occupying the two half bands. So, when both half bands are considered in this way, the magnetization is

$$M_v = \mu_B \cdot 2\Delta N = 2N(E_F)\mu_B{}^2 H \qquad (8\text{-}118)$$

and the magnetization per unit volume is

$$M = \frac{M_v}{V} = \frac{2N(E_F)\mu_B{}^2 H}{V} \qquad (8\text{-}119)$$

This gives the susceptibility as

$$\chi_p = \frac{M}{H} = \frac{2N(E_F)\mu^2{}_B}{V} \qquad (8\text{-}120)$$

Equation 5-21 is used to obtain

$$N(E_F) = \frac{2\pi V}{h^3} (2m)^{3/2} E_F{}^{1/2}$$

or, more conveniently,

$$\frac{N(E_F)}{V} = \frac{2\pi}{h^3}(2m)^{3/2} E_F{}^{1/2} = \frac{2\pi}{h^3} 2\sqrt{2}\,(m)^{3/2} E_F{}^{1/2} \qquad (8\text{-}121)$$

The Fermi level for unit volume is given, by setting $V = 1$, in Equation 5-24 to obtain

$$E_F = \frac{h^2}{8m} \left[\frac{3N}{\pi}\right]^{2/3} \qquad (8\text{-}122)$$

From this,

$$E_F^{1/2} = \frac{h}{2\sqrt{2}\, m^{1/2}} \left[\frac{3N}{\pi}\right]^{1/3} \tag{8-123}$$

The substitution of Equation 8-123 into Equation 8-121 gives

$$\frac{N(E_F)}{V} = \frac{2\pi}{h^3}\, 2\sqrt{2}\, m^{3/2} \frac{h}{2\sqrt{2}\, m^{1/2}} \left[\frac{3N}{\pi}\right]^{1/3}$$

which reduces to

$$\frac{N(E_F)}{V} = \frac{2\pi m}{h^2} \left[\frac{3N}{\pi}\right]^{1/3} \tag{8-124}$$

Equation 8-124 now is substituted into Equation 8-120:

$$\chi_p = \frac{M}{H} = \frac{4\pi m}{h^2} \left[\frac{3N}{\pi}\right]^{1/3} \mu_B^2 \tag{8-125}$$

The numerator and denominator of Equation 8-125 are multiplied by E_F, or its equivalent (Equation 8-122) to obtain

$$\chi_p = \frac{4\pi m}{h^2} \left[\frac{3N}{\pi}\right]^{1/3} \mu_B^2 \frac{h^2 \left[\frac{3N}{\pi}\right]^{2/3}}{8mE_F} = \frac{3N\mu_B^2}{2E_F} = \frac{3N\mu_B^2}{2k_BT_F} \tag{8-126}$$

This result may be verified in a simple way starting with Equation 8-117. This gives an expression for the magnetization as

$$M = \Delta N\mu_B = \mu_B^2 HN(E_F) \tag{8-127}$$

Figure 5-5c Chapter 5, Volume I is used to approximate $N(E_F)$: $N = 2N(E_F,0)E_F/3$, since the curve is parabolic. This gives $N(E_F) \simeq N(E_F,0) = 3N/2E_F$. Thus, Equation 8-127 becomes

$$M = \mu_B^2 H \frac{3N}{2E_F}$$

and

$$\chi_p = \frac{M}{H} = \frac{3N\mu_B^2}{2E_F} = \frac{3N\mu_B^2}{2k_BT_F} \tag{8-128}$$

Both equations for the Pauli paramagnetic susceptibility of the valence electrons (Equations 8-126 and 8-128) are identical and larger than that for the Landau diamagnetism of these electrons by a factor of three (see Equation 8-34). All are virtually constant.

Equations 8-126 and 8-128 both were based on the assumption of the invariance of E_F with temperature. Corrections may be made for the weak temperature dependence of E_F by use of the following modification

$$\chi_p = \frac{3N\mu_B^2}{2E_F(T)} \tag{8-128a}$$

where $E_F(T)$ is given by Equation 5-26.

The important magnetic effects discussed thus far in this chapter may be summarized as follows:

1. Based upon Larmor precession for ion cores:

$$M = -\frac{NZe^2H}{4mc^2}\sum_i r_i^2 \qquad (8\text{-}31)$$

$$\chi_D = -\frac{NZe^2}{4mc^2}\sum_i r_i^2 \qquad (8\text{-}32)$$

2. Based upon the Langevin approach for ion cores:

$$M = -\frac{NZe^2\overline{r}^2H}{6mc^2} \qquad (8\text{-}33)$$

$$\chi_D = -\frac{NZe^2\overline{r}^2}{6mc^2} \qquad (8\text{-}33)$$

3. Based upon the Landau approach for valence electrons:

$$M = -\frac{N\mu_B^2H}{2k_BT_F} \qquad (8\text{-}34)$$

$$\chi_D = -\frac{N\mu_B^2}{2k_BT_F} \qquad (8\text{-}34)$$

4. Based upon the Langevin approach for freely rotating dipoles:

$$M = \frac{N\mu_M^2H}{3k_BT} \qquad (8\text{-}50)$$

$$\chi_p = \frac{N\mu_M^2}{3k_BT} \qquad (8\text{-}51)$$

5. Based upon the Landé spectrographic splitting factor:

$$M = \frac{Ng^2J^2\mu_B^2H}{k_BT} = \frac{N\mu_B^2H}{k_BT} \quad \text{(for spin alone)} \qquad (8\text{-}84)$$

$$\chi_p = \frac{Ng^2J^2\mu_B^2}{k_BT} = \frac{N\mu_B^2}{k_BT} \quad \text{(for spin alone)} \qquad (8\text{-}85)$$

6. Based upon the total quantum number, J:

$$M = \frac{Ng^2J(J+1)}{3k_BT}\mu_B^2H = \frac{N\mu_B^2H}{k_BT} \quad \text{(for spin alone)}$$

(8-101)

$$\chi_p = \frac{Ng^2J(J+1)}{3k_BT}\mu_B^2 = \frac{N\mu_B^2}{k_BT} \quad \text{(for spin alone)}$$

(8-102)

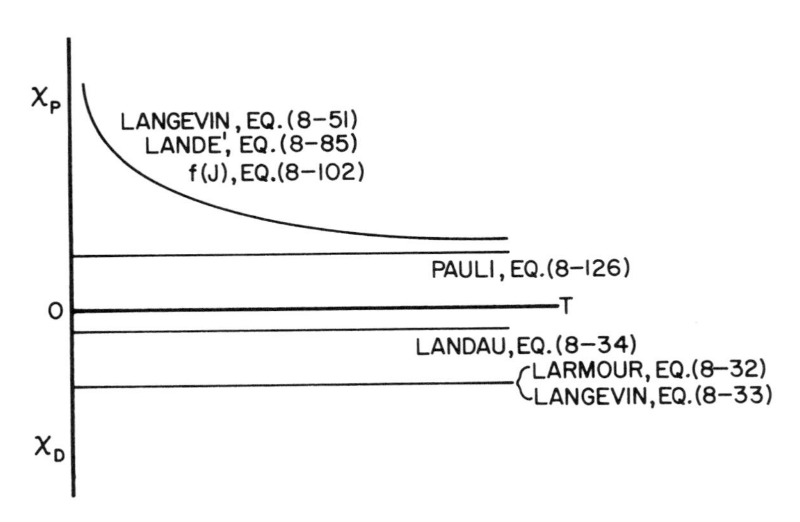

FIGURE 8-17. Schematic diagram of paramagnetic and diamagnetic susceptibilities as functions of temperature. (Based on Kittel, C., *Introduction to Solid State Physics*, 3rd ed., John Wiley & Sons, New York, 1966, 428. With permission.)

7. Based upon the Pauli approach for valence electrons:

$$M = \frac{3N\mu_B^2 H}{2k_B T_F} \tag{8-128}$$

$$\chi_p = \frac{3N\mu_B^2}{2k_B T_F} \tag{8-128}$$

These relationships are shown schematically in Figure 8-17.
The magnetic effects of atomic nuclei are considered in the next section.

8.4. NUCLEAR PARAMAGNETISM

Hyperfine structures of spectral lines, very close to the wavelengths of the original lines, result from small magnetic moments which are associated with the nuclei of atoms. This arises from proton spin. The nuclear spin quantum number is designated by I. The nuclear angular momentum resulting from proton spin is given in a way analogous to Equation 8-56, by

$$L_I = \frac{h}{2\pi}\,[I(I+1)]^{1/2} \tag{8-129}$$

The nuclear spin quantum number may take on $2I+1$ values, m_I, in a magnetic field in a way analogous to $m\ell$ (Section 8.3.2). Similarly, the nuclear magnetic moment is given by

$$\mu_I = \frac{e}{2M_p c}\cdot\frac{h}{2\pi}\,[I(I+1)]^{1/2} = \mu_N\,[I(I+1)]^{1/2} \tag{8-130}$$

in which M_p and e are the mass and charge on a proton, respectively. This equation corresponds to Equation 8-57. The ratio of these equations is

$$\frac{\mu_B}{\mu_N} = \frac{\frac{\hbar e}{2mc}}{\frac{\hbar e}{2M_pc}} = \frac{M_p}{m} \quad (8\text{-}131)$$

or, $\mu_N = 1/1836\ \mu_B$ and is known as a nuclear magneton. It has a value of 5.05×10^{-24} erg/Oe. Direct measurement of nuclear magnetic behavior, thus, is possible only for atoms or molecules where the electron magnetic moment is negligibly small.

The experimental data show that the nuclear magnetic moment is not always proportional to the angular momentum as defined by Equation 8-130. The nuclear "g factor" is so defined as to take this anomalous behavior into account by means of the relation

$$g_N = \frac{\mu_N'/\mu_N}{I/\hbar} \quad (8\text{-}132)$$

because μ_N'/μ_N varies from about -1 to $+4$ in a nonintegral way. Under these conditions, g_N may range from about 0.1 to 5.6, depending upon the given nucleus. No theory, comparable to that given for electrons, can as yet account for such variations.

This behavior is taken into consideration in the expression for nuclear susceptibility first by modifying Equation 8-130 to read

$$\mu_N' = g_N\mu_N [I(I+1)]^{1/2} \quad (8\text{-}133)$$

This, in turn, is used to express the nuclear susceptibility as

$$\chi_N = \frac{N(\mu_N')^2}{3k_BT} \quad (8\text{-}134)$$

The large difference between μ_N and μ_B accounts for the differences in the magnitudes of magnetic effects resulting from protons and electrons.

The nucleus precesses in a magnetic field in a way similar to that of electrons (see Equation 8-26). The frequency is given, by analogy with Equation 8-77b, as

$$\omega = -g_N \frac{eH}{2M_pc} = -\frac{g_N\mu_N}{\hbar} H \quad (8\text{-}135)$$

This frequency is smaller than that for electrons by a factor of 1/1836; it is in the range of radio frequencies. Thus, for relatively weak fields and suitable frequencies, a coupling of nuclear paramagnetic resonance and electron spin resonance (Section 8.3.2.1) can occur simultaneously. This makes it possible to determine the ways in which nuclei and electrons affect each other; this is employed to measure nuclear spins. If the field is too strong, this coupling does not exist because electron and nuclear spin orientations become independently oriented and no longer are coupled.

8.5. ADIABATIC PARAMAGNETIC COOLING

Very low temperatures, quite close to the absolute zero, have been obtained by the use of adiabatic demagnetization. It is based upon the fact that the entropy of the electrons of dipolar ions is much greater than the entropy of the lattice. It will be recalled that the electron energy is high and is virtually unaffected by temperature while the lattice internal energy approaches a constant value which, comparatively, is very small (see Section 4.2, Chapter 4, Volume I). These are different by a factor of about 200. The isothermal application of a magnetic field to the solid, thus, has its greatest effect upon the electrons. The lattice is very regular at these low temperatures. However, the electrons show random spin behavior prior to the application of the field. The field "lines up," or orders, a large percentage of the electron spins as well as some of the ions. This decreases their entropy. The large decrease in the entropy of the electrons corresponds to an ordered behavior to be expected at a significantly lower temperature. The decrease in the entropy of the ions is very small; they previously were well aligned. Effectively, the electrons are at a lower temperature and energy than is the lattice. The thermal energy drains from the lattice to the electrons, leaving it at a lower temperature. When the magnetic field is turned off, the electrons resume random behavior corresponding to their newly increased energies. Prior to the time that the field is turned off, the specimen is insulated to prevent thermal exchange with its surroundings. This process may be repeated. Temperatures of the order of 10^{-2} to 10^{-3} K have been achieved by this means.

This behavior may be clarified by referring to the thermodynamics involved. Starting with the equation for the Gibbs free energy

$$F = E - TS$$

in which E and S are the internal energy and entropy, respectively. Since the internal energy of the ions is small compared to that of the electrons, it will be neglected and the approximation is made that

$$E \simeq MH$$

and the Gibbs free energy becomes

$$F \simeq MH - TS$$

or, in differential form,

$$dF = MdH - SdT$$

Since dF is a function of dH and dT, it can be written as

$$dF = \left[\frac{\partial F}{\partial H}\right]_T dH + \left[\frac{\partial F}{\partial T}\right]_H dT$$

Also, $M = (\partial F/\partial H)_T$ and $S = (\partial F/\partial T)_H$. Thus, by differentiation,

$$\left[\frac{\partial M}{\partial T}\right]_H = \left\{\frac{\partial}{\partial T}\left[\frac{\partial F}{\partial H}\right]_T\right\}_H = \frac{\partial^2 F}{\partial H \partial T}$$

and

$$\left[\frac{\partial S}{\partial H}\right]_T = \left\{\frac{\partial}{\partial H}\left[\frac{\partial F}{\partial T}\right]_H\right\}_T = \frac{\partial^2 F}{\partial H \partial T}$$

So, being equal to the same quantity, it follows that

$$\left[\frac{\partial M}{\partial T}\right]_H = \left[\frac{\partial S}{\partial H}\right]_T$$

Upon integration, the change in entropy which occurs between the field-on and field-off conditions is

$$\Delta S = \int_0^H \left[\frac{\partial M}{\partial T}\right]_H dH$$

If the material is paramagnetic and follows the Curie law (Equations 8-51 and 8-52, then

$$M = \frac{CH}{T}$$

$$C = N\mu_M^2/3k_B$$

and

$$\frac{\partial M}{\partial T} = -\frac{CH}{T^2}$$

Thus, the change in entropy is

$$\Delta S = \int_0^H \left[-\frac{CH}{T^2}\right]_H dH = -C/2 \cdot H^2/T^2 \qquad (8\text{-}136)$$

Under initial experimental conditions H and T are about 10^4 gauss and 1 K, respectively. Therefore, the decrease in entropy is large. It represents a degree of order characteristic of a significantly lower temperature.

Another way of picturing adiabatic cooling is to consider the total entropy of the solid, S_s, as

$$S_s = S_L + S_e$$

where S_L and S_e are the entropies of the lattice and electrons, respectively. This may be expressed in terms of the respective heat contents as

$$S_s = \frac{Q_L}{T_2} + \frac{Q_e}{T_2}$$

$$Q_L \ll Q_e$$

The entropy of the electrons greatly decreases when the field is applied. This means that the entropy of the electrons now is $Q'_e \ll O_e$ and that the electrons now are at a temperature $T_1 < T_2$. The thermal energy, thus, will flow from the lattice to the electrons. The entropy of the solid, at the time of demagnetization, is

$$S_s = \frac{Q'_L}{T_1} + \frac{Q'_e}{T_1}$$

This is shown schematically in Figure 8-18. The lower temperature, T_1, will be maintained if the cooled substance is insulated from its surroundings.

The effects of adiabatic demagnetization upon the entropy and temperature of a paramagnetic solid are indicated in Figure 8-19 as functions of time.

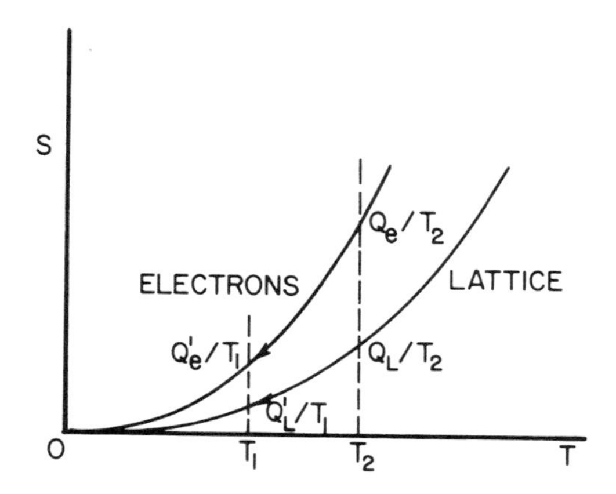

FIGURE 8-18. Changes in electron and lattice entropies caused by the adiabatic application of a magnetic field at T_2. The decrease in electron entropy to Q'_ℓ/T_1 cools the lattice to T_1.

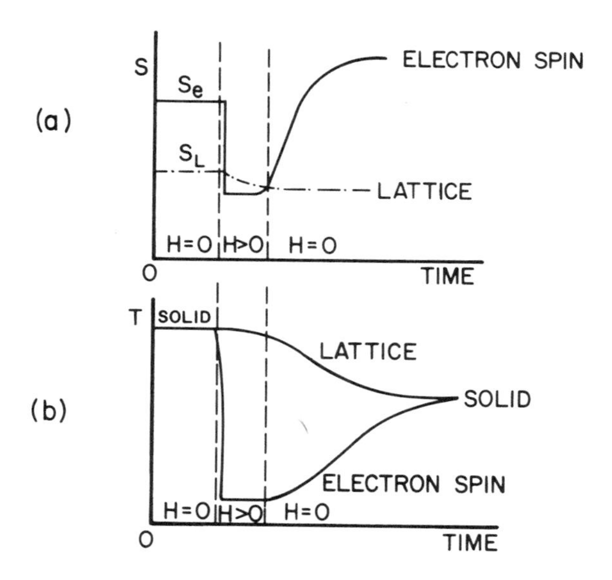

FIGURE 8-19. (a) Changes in entropy, (b) in temperature during adiabatic magnetic cooling. (Modified from Kittel, C., *Introduction to Solid State Physics,* 3rd ed., John Wiley & Sons, 1966, 441.)

This type of cooling is produced in practice in the following way. An evacuated container holding a paramagnetic salt is placed in a liquid helium bath. The vapor pressure of the He is lowered by pumping until the He reaches a temperature of about 1 K. Low-pressure He gas is admitted into the previously evacuated container to serve as a medium for the transfer of heat from the salt in the container to the liquid He bath. A magnetic field is then applied to the apparatus, magnetizing the salt. The He gas removes the heat generated by the magnetization so that this is accomplished with virtually no change in the temperature of the salt. The He gas then is removed so that the subsequent demagnetization occurs under essentially adiabatic conditions. The decrease in entropy and the corresponding decrease in temperature are obtained in this way.

8.6. PARAMAGNETIC SOLIDS

The transition and rare-earth elements and their compounds show the highest degree of paramagnetism. As previously noted (Section 3.11.2, Chapter 3, Volume I) these atoms have unfilled inner electron states. These are only partially involved in bonding, and to a much smaller degree than valence electrons (see Section 10.7.3, Chapter 10, Volume III) and, thus, the remaining unpaired electrons give rise to magnetic moments.

As described in the above-noted section, the atoms of the first transition series have incomplete 3d states. The 4d states of the next transition series, Y through Pd, are partly filled. The rare-earth elements have unfilled 4f states. The next series ranges from Hf to Pt; these have less than 10 5d states. The last series begins with Pa and consists of atoms with incompleted 5f and 6d levels.

Compounds such as ionic salts, oxides, sulphates, and carbonates of the first transition group and the rare earth elements have paramagnetic susceptibilities which vary with temperature as given by a modification of Equation 9-8, the Curie-Weiss law, in the form

$$\chi_P = \frac{C}{T - \theta} - \alpha \qquad (8\text{-}137)$$

where θ is the Weiss constant and α usually accounts for the small diamagnetic effects of the filled electron shells. The parameter α is independent of temperature. The value of θ varies widely, being nearly zero for dilute compounds of transition metals; it may be in the range of 20 to 50 K for salts with higher concentrations of these ions. The experimentally obtained data frequently are expressed in terms of the effective number of Bohr magnetons. This parameter can be simplified by means of substituting Equation 8-104a in Equation 8-102.

$$\chi_P = \frac{Ng^2J(J+1)\mu_B^2}{3k_BT} = \frac{C}{T} \qquad (8\text{-}102)$$

$$n_B^2 = g^2J(J+1) \qquad (8\text{-}104a)$$

The substitution gives, where n_B is the effective number of Bohr magnetons per ion (see Equations 8-89 and 8-90),

$$\frac{Nn_B^2\mu_B^2}{3k_B} = C$$

and

$$n_B\mu_B = \left[\frac{3k_BC}{N}\right]^{1/2}$$

$$n_B^2 = \frac{3k_BC}{N\mu_B^2} \qquad (8\text{-}138)$$

The effective number of Bohr magnetons, determined in this way, is used to describe the magnetic behavior of many substances.

In crystalline substances, an ion at a given time will be in an allowed quantum state. As its neighbors are influenced by phonons, the given ion will be affected by the changes in its electric and magnetic environment induced by the neighboring ions. Umklapp processes can occur in which the energy states of some electrons are changed. This happens repeatedly throughout the lattice. At any given time, a specific percentage of the ions will be in any one of the allowed quantum states. At thermal equilibrium, a dynamic equilibrium of the percentages of ions in allowed states is established which determines n_B.

The energies will be changed when a field is applied; those ions with dipole moments opposite to the field will be decreased, those with dipole moments parallel will be increased. In a manner similar to that described in Section 8.3.3, a net number of ions whose dipole moments are parallel to the field will exist. This causes the paramagnetic behavior. With increasing temperature, the umklapp processes cause the net number of dipole moments parallel to the field to diminish. This results in decreasing the susceptibility with increasing temperature.

The umklapp interactions with the phonons are necessary for pronounced paramagnetism because they provide the means for the ions to change their orientations with respect to the field. These interactions are not required to change the energies of the ions greatly. Their role is to provide sufficient numbers of energy and orientation transitions to achieve the distribution required by thermal equilibrium.

In the case of some transition elements, the ions are affected by interactions with their neighbors. This has an important effect upon the orbital momenta of their electrons. J is strongly affected when this occurs. The component due to orbital motion is absent (see Section 8.6.2).

8.6.1. Compounds of Rare Earth Elements

The susceptibilities of these compounds obey the Curie-Weiss law (see Equation 9-8 in Chapter 9) at and above room temperature. The value of n_B, the effective number of Bohr magnetons per ion, is essentially constant for a given ion, Equation 8-138. This verifies Equations 8-102 and 8-104a. Some compounds, such as those of samarium and europium, do not follow these relationships. It also should be noted that these substances do not agree closely with the Curie-Weiss law.

The accuracy of prediction of Equation 8-104a, with the exception of the two cases noted above, is good because the assumption of high occupancy of the lowest J state is valid. This is shown in Table 8-3.

It is necessary to refer to Hund's rules (Equation 8-59) to obtain the proper values of J. The discrepancies shown by Eu and Sm arise because their lowest J states are very close together ($\simeq sk_BT$). The calculation for the lowest value of J for Eu^{+3} is zero ($J = L - S = 3 - 3 = 0$). Higher J levels, rather than the lowest, are occupied in the neighborhood of room temperature and Equation 8-104a gives erroneous results. As a consequence, calculations for n_B are incorrect.

The other ions of this class which have calculated values for n_B which are in good agreement with the experimental values behave in this way because the electrons contributing to the magnetic effects are not significantly affected by the crystal fields (section 8.6.2). These electrons remain closely associated with their respective ions, permitting the use of Equations 8-102 and 8-104a.

8.6.2. Compounds of the First Transition Series

Many of the compounds of the iron-group elements agree with the Curie-Weiss relationship. As before, the effective number of Bohr magnetons per ion can be determined from Equation 8-138. The value of n_B for a given ion is approximately the same in most compounds of that ion.

Equation 8-102 cannot explain the behavior of these iron-group ions as it does that of the rare-earth ions. This results from an effect known as crystal-field splitting. The

Table 8-3
EFFECTIVE MAGNETON NUMBERS FOR TRIVALENT LANTHANIDE GROUP IONS[2] (NEAR ROOM TEMPERATURE)

Ion	Configuration	Basic Level	Calculated $n_B = g[J(J+1)]^{1/2}$	Approximate experimental n_B
Ce^{3+}	$4f^1\ 5s^2\ p^6$	$^2F_{5/2}$	2.54	2.4
Pr^{3+}	$4f^2\ 5s^2\ p^6$	$^3H_4$	3.58	3.5
Nd^{3+}	$4f^3\ 5s^2\ p^6$	$^4I_{9/2}$	3.62	3.5
Pm^{3+}	$4f^2\ 5s^2\ p^6$	$^5I_4$	2.68	—
Sm^{3+}	$4f^5\ 5s^2\ p^6$	$^6H_{5/2}$	0.84	1.5
Eu^{3+}	$4f^6\ 5s^2\ p^6$	$^6F_0$	0	3.4
Gd^{3+}	$4f^7\ 5s^2\ p^6$	$^8S_{7/2}$	7.94	8.0
Tb^{3+}	$4f^8\ 5s^2\ p^6$	$^7F_6$	9.72	9.5
Dy^{3+}	$4f^9\ 5s^2\ p^6$	$^6H_{15/2}$	10.63	10.6
Ho^{3+}	$4f^{10}\ 5s^2\ p^6$	$^5I_8$	10.60	10.4
Er^{3+}	$4f^{11}\ 5s^2\ p^6$	$^4I_{15/2}$	9.59	9.5
Tm^{3+}	$4f^{12}\ 5s^2\ p^6$	$^3H_6$	7.57	7.3
Yb^{3+}	$4f^{13}\ 5s^2\ p^6$	$^2F_{7/2}$	4.54	4.5

[2] Also see Reference 6 (Table 3.3 a).

From Kittel, C., *Introduction to Solid State Physics*, 5th ed., John Wiley & Sons, New York, 1976, 437. With permission.

3d levels responsible for the paramagnetic effects of the iron-group ions lie at the outside of the ion. The 4f levels responsible for the properties of the rare-earths are buried, or screened, beneath the 5s and 5p levels and remain unaffected (see Table 8-3). As noted in Section 8.6, the outer 3d levels are affected by the relatively strong electric fields induced by neighboring ions, while the screened 4f levels are not. The motions of the d electrons are influenced by reactions with electrons of surrounding ions. Their wave functions overlap and some degree of covalency can occur. Stoner (1929) lumped these effects as being equivalent to those of a virtual, or equivalent, internal field known as a crystal field. The orbital electron moments are considered to be oriented with respect to the crystal field so that they cannot react to external fields. The orbital momentum is said to be quenched by the crystal field. Thus far, it has not been possible to explain the magnitudes of crystal fields. Electron spin is not directly affected by the crystal field and is free to be aligned with external fields.

Therefore, the crystal field surrounding the 3d levels has important effects. It results in the splitting up of the L-S coupling, since L = 0, and the parameter J, as such, is no longer operative, J = S. Thus, the paramagnetic behavior of the iron group ions arises from spin alone. On this basis, Equations 8-102 and 8-104a become

$$\chi_p = \frac{Ng^2 S(S+1)\mu_B^2}{3k_B T} \qquad (8\text{-}139)$$

and

$$n_B^2 = g^2 S(S+1) \qquad (8\text{-}140)$$

These equations assume that spin alone is operative and that the orbital momentum is entirely quenched by the crystal field. Data obtained in both ways, using both J and S, are given in Table 8-4.

Table 8-4
EFFECTIVE MAGNETON NUMBERS FOR IRON GROUP IONS

Ion	Configuration	Basic level	Calculated $n_B = g[J(J+1)]^{1/2}$	Calculated $n_B = g[S(S+1)]^{1/2}$	n_B (exp)[a]
Ti^{+3}, V^{+4}	$3d^1$	$^2D_{3/2}$	1.55	1.73	1.8
V^{+3}	$3d^2$	$^3F_2$	1.63	2.83	2.8
Cr^{+3}, V^{+2}	$3d^3$	$^4F_{3/2}$	0.77	3.87	3.8
Mn^{+3}, Cr^{+2}	$3d^4$	$^5D_0$	0	4.90	4.9
Fe^{+3}, Mn^{+2}	$3d^5$	$^6S_{5/2}$	5.92	5.92	5.9
Fe^{+2}	$3d^6$	$5D_4$	6.70	4.90	5.4
Co^{+2}	$3d^7$	$^4F_{9/2}$	6.63	3.87	4.8
Ni^{+2}	$3d^8$	$^3F_4$	5.59	2.83	3.2
Cu^{+2}	$3d^9$	$^2D_{5/2}$	3.55	1.73	1.9

[a] Representative values.

From Kittel, C., *Introduction to Solid State Physics*, 5th ed., John Wiley & Sons, New York, 1976, 438. With permission.

It would appear from the calculations as compared with the experimental data that the orbital moments are almost completely absent. The orbital moments are nearly totally quenched. Spin is the operative factor as evidenced by the high degree of agreement of n_B based upon spin alone with the corresponding experimental values.

8.6.3. Compounds of the Later Transition Ions

The Curie-Weiss law is not usually obeyed by the palladium and platinum series of ions. Their susceptibilities generally are more complex and are smaller than for their iron-group equivalents. The calculated values of χ_p and n_B (Equations 8-139 and 8-140 are very much larger than experimental values.

As is the case for the first transition ions, the d levels of these elements are affected by the crystal field, but their spin-orbit couplings can be large. This is considerably different from the crystal-field splitting which affects the iron-group elements, and it accounts for some of the differences between compounds of these two groups of elements. It also has been considered that the crystal field may be responsible for the promotion of some of the electrons in incompleted shells so that they enter into bonding (see Section 10.7.3, Chapter 10, Volume III). This would suppress some of the spin moments and affect the susceptibility.

8.6.4. Pure Transition Elements

A large span of magnetic variation is manifested by the metals of the first transition group. Unlike the rare-earth ions, the d levels of this group are not screened by completed, outer electron levels. The effects of the crystal field are much more pronounced on these ions in the metallic state than when they are associated with other ions in compounds. In addition, some of the d states of a given atom are not completely fixed with respect to a given ion and these are itinerant throughout the lattice (see Section 9.9, Chapter 9). The wide range in the properties of those elements in the first transition series, starting with chromium and ending with nickel, results from overlapping wave functions and exchange reactions (see Section 9.5, Chapter 9). This leads to the ferromagnetism of iron, cobalt, and nickel. These elements are paramagnetic above their Curie temperatures. Above these temperatures, they show paramagnetic behaviors and approximately follow the Curie-Weiss law. The Curie constants of these elements have not yet been explained on the basis of their ionic configurations.

The nonferromagnetic members of this group of elements show positive susceptibilities which vary extremely weakly with temperature. This almost negligible temperature dependence arises from the conduction electrons. On a comparative basis, the susceptibilities of these elements are considerably larger than those of normal elements, but are small compared to those of ionic compounds. Alloys with elements of the first transition series can have strongly temperature-dependent, positive susceptibilities. This leads to the idea that the electrons responsible for this must be closely associated with their respective ions. Other alloys with virtually temperature-independent susceptibilities are explained by the conduction electrons, which are not associated with any given ion.

As previously noted, the electrons responsible for the paramagnetic properties of the rare earths are shielded by outer levels and are not significantly affected by the crystal fields. Most of these elements obey the Curie-Weiss law. Exchange interactions between the ions can account for the magnitudes of the Weiss constants. Antiferromagnetic or ferromagnetic behavior results from these interactions at temperatures below those of the Weiss constants.

The members of the later transition groups are not ferromagnetic, but behave in a way similar to those of the first transition series. Their paramagnetic susceptibilities are only slightly temperature dependent. This is explained on the basis of the conduction electrons.

8.7. PROBLEMS

1. Using data in Table 8-1, calculate the atomic and molar magnetic susceptibilities of Al, Mg, Ti, Mn, and Cu.
2. Employ Equation 8-33 and data in Table 10-3, Chapter 10, Volume II, to compute the mass magnetic susceptibilities of Al, Mg, Ti, Mn, and Cu. Compare the results with the data in Table 8-1. Explain any significant differences.
3. Discuss the effects of temperature upon the diamagnetic susceptibility of a solid.
4. Show that $\overline{r^2} = 2/3\ r^2$ so that Equation 8-33 is possible.
5. Show that the Landé spectroscopic splitting factor equals 2.0 for spin and 1.0 for angular orbital momentum.
6. Discuss the reasons why the Landau and Pauli effects must be independent of temperature while other sources of diamagnetism and paramagnetism are temperature dependent.
7. Discuss the bases for the differences between Equations 8-34 and 8-128.
8. Compare the differences between Equations 8-33 and 8-112a in terms of the types of materials to which they apply.
9. Draw a diagram for the Russell-Saunders coupling for the $s_i - s_j$ and $\ell_i - \ell_j$ couplings of two electrons in a magnetic field.
10. Calculate J for 7, 8, 9, and 10 3d electrons, respectively.
11. Draw the vector diagrams for (a) $\ell = 2$ and $s = 1/2$; and (b) $\ell = 3$ and $s = 1/2$. Show their values and those of the resultants.
12. Draw a sketch of the Zeeman effect for $J = 5/2$. Label the levels and the gaps.
13. Discuss all of the magnetic effects which apply to the following diamagnetic materials: (a) a metal; (b) a metal-halide salt; and (c) a solid organic material.
14. Discuss all of the magnetic effects which apply to a paramagnetic metal.
15. Approximate the magnetic susceptibility of sodium at room temperature. Include all effects. Compare the results with the data in Table 8-1.
16. Discuss the effects of temperature and a magnetic field upon the energy and entropy of the electrons in a metal. Make a comparison with the ions in a metal for the same variables.
17. Why should paramagnetic cooling be conducted adiabatically considering that the initial temperature is about 1 K?

8.8. REFERENCES

1. **Dekker, A. J.,** *Solid State Physics*, Prentice-Hall, Englewood Cliffs, N.J., 1957.
2. **Bates, L. F.,** *Modern Magnetism*, MIT Press, Cambridge, 1967.
3. **White, H. E.,** *Introduction to Atomic Spectra*, McGraw-Hill, New York, 1934.
4. **Kittel, C.,** *Introduction to Solid State Physics*, John Wiley & Sons, New York, 1966.
5. **Stanley, J. K.,** *Electrical and Magnetic Properties of Metals*, American Society for Metals, Metals Park, Ohio, 1963.
6. **Martin, D. H.,** *Magnetism in Solids*, MIT Press, Cambridge, 1967.
7. **Levy, R. A.,** *Principles of Solid State Physics*, Academic Press, New York, 1968.
8. **Sproull, R. L.,** *Modern Physics*, 2nd ed., John Wiley & Sons, New York, 1963.
9. **Slater, J. C.,** *Quantum Theory of Matter*, McGraw-Hill, New York, 1951.
10. **Chikazumi, S.,** *Physics of Magnetism*, John Wiley & Sons, New York, 1964.
11. **Morrish, A. H.,** *The Physical Principles of Magnetism*, John Wiley & Sons, New York, 1965.
12. **Martin, D. M.,** *Magnetism in Solids*, MIT Press, Cambridge, 1967.

Chapter 9

FERROMAGNETISM

In contrast to diamagnetic and paramagnetic materials which show a weak response to the presence of an applied magnetic field ferromagnetic materials show a very strong response. The most well-known ferromagnetic metals are Fe, Co, Ni, and Gd. Other rare-earth elements are ferromagnetic at very low temperatures. In addition, many alloys and compounds show this behavior.

Ferromagnetic materials show two important characteristics. First, relatively large magnetizations may be induced in them spontaneously or by comparatively low external fields. Second, these materials may retain some magnetization when the field is turned off. Both of these result from the spontaneous alignment of the magnetic moments in the material; a cooperative effect is present.

These properties lead to the domain or Weiss molecular field theory in which it was postulated that small, spontaneously magnetized volumes, or domains, exist in the substance. Each domain in the substance consists of a volume of the substance in which the magnetic moments are all aligned in the same direction. The magnetization of a specimen is determined by the resultant of all of the vectors of all of the domains.

The presence of these domains can be demonstrated by applying a colloidal suspension of magnetic particles to the polished surface of a magnetic material. The particles are attracted to the places of strong magnetization at domain boundaries and, thus, outline the intersections of the domains with the polished surfaces. The patterns thus formed are called "Bitter patterns". When an external field is applied to a magnetic specimen, which has been prepared in this way, the patterns change. This makes it possible to observe the behavior of the domains.

9.1. THE MOLECULAR FIELD MODEL

In contrast to paramagnetic materials, ferromagnetic substances may be maximumly magnetized (reach saturation magnetization) in small, or negligible, fields at normal temperatures. Weiss hypothesized that the line-up of magnetic moments within a domain resulted from a spontaneous, cooperative interaction between the atoms of which it is composed. The Weiss, or molecular, field is regarded as the equivalent of an internal magnetic field acting on and aligning the spins of the electrons. It was assumed that such a molecular field could be given by the proportional relationship

$$H_I = \lambda M \qquad (9\text{-}1)$$

for spontaneous magnetization in which λ is the molecular field constant. It should be noted that H_I is about 10^6 to 10^7 Oe for iron, a field strength of greater magnitude than has as yet been produced experimentally. Then, the total field may be given by

$$H_T = H + \lambda M \qquad (9\text{-}2)$$

where H is the applied field.

In the case of paramagnetic materials it was shown, using the notation $B_J(x)$ instead of $B_J(a)$, that

$$M = Ng\mu_B J B_J(x) \qquad (8\text{-}112)$$

where

$$a = x = \frac{Jg\mu_B H}{k_B T} \tag{9-3}$$

This equation may be applied to ferromagnetic materials by realizing that the field here is H_T instead of just H. Thus, Equation 9-3 may be written for this case where

$$x = \frac{Jg\mu_B (H + \lambda M)}{k_B T} \tag{9-4}$$

The value of H is zero for spontaneous magnetization, so Equation 9-4 becomes

$$x = \frac{Jg\mu_B \lambda M}{k_B T} \tag{9-4a}$$

Solving Equation 9-4a for M gives

$$M = \frac{x k_B T}{Jg\mu_B \lambda} \tag{9-5}$$

The simultaneous solutions of Equations 8-112 and 9-5 give the spontaneous magnetization at a given temperature, as shown in Figure 9-1. This may be done graphically as shown in the figure.

The amount of spontaneous magnetization increases as the quantity $(T_c - T)$ increases. The parameter T_c is that temperature at which the spontaneous magnetization is zero; it is known as the Curie Temperature. The Curie temperature frequently is designated by θ. Thus, the lower the temperature of the substance is with respect to T_c, the greater will be the spontaneous magnetization. This increase slowly approaches a limit as T decreases because of the nature of Equation 8-112 which reflects a maximum spin alignment at 0 K. For $T > T_c$, spontaneous magnetization is absent; the material becomes paramagnetic. This behavior of magnetization up to T_c, as a function of temperature, is shown in Figure 9-2.

The properties for $T > T_c$ can be obtained by again referring to paramagnetic behavior. It was found that

$$M = \frac{Ng^2 J(J + 1)\mu_B^2 H}{3k_B T} \tag{8-101}$$

This also can be applied to ferromagnetic behavior by using the same substitution as was employed for Equation 9-4. Then Equation 8-101 becomes

$$M = \frac{Ng^2 J(J + 1)\mu_B^2 (H + \lambda M)}{3k_B T} \tag{9-6}$$

This can be simplified to

$$M = \frac{C}{T}(H + \lambda M)$$

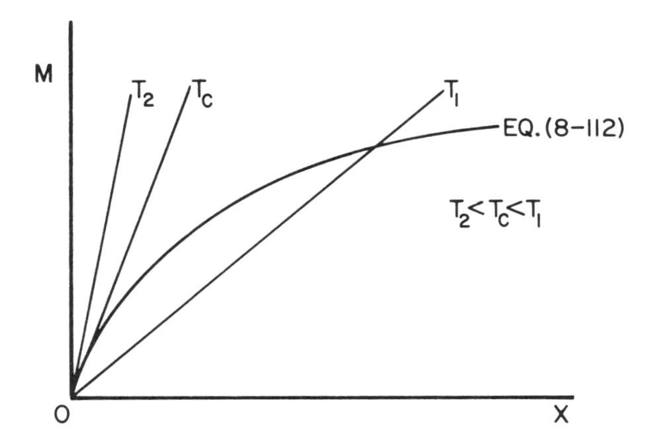

FIGURE 9-1. Simultaneous solutions of Equations 8-112 and 9-5.

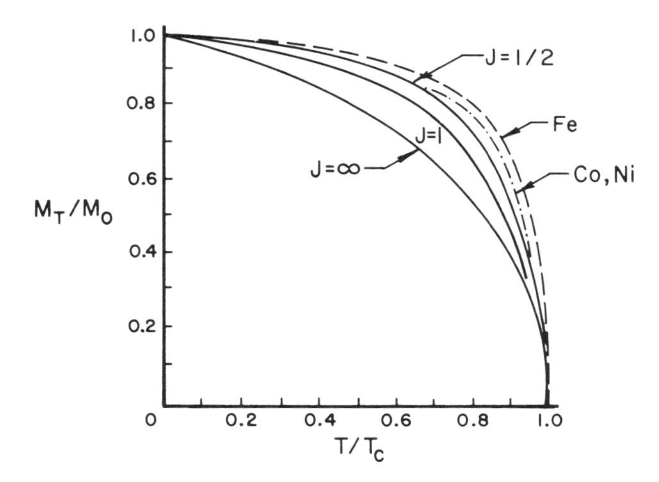

FIGURE 9-2. Relative saturation magnetization normalized with respect to the Curie temperature with experimental data for Fe, Co, and Ni. (After Bozorth, R. M., *Ferromagnetism*, Van Nostrand, New York, 1951, 431. With permission.)

$$C = \frac{Ng^2 J(J + 1)\mu_B^2}{3k_B} \tag{9-7}$$

where all of the constants are included in C. Equation 9-7 may be rewritten as

$$MT = CH + C\lambda M \tag{9-7a}$$

or as

$$M(T - C\lambda) = CH$$

Then, for $T > T_c$,

$$\chi_p = \frac{M}{H} = \frac{C}{T - C\lambda} = \frac{C}{T - T_c} \tag{9-8}$$

This is the Curie-Weiss law for paramagnetic behavior of ferromagnetic materials at temperatures above T_c. The Curie temperature is proportional to the molecular field constant. Using Equations 9-7 and 9-8, for H = 0,

$$T_c = C\lambda = \frac{Ng^2 J(J+1)\mu_B^2}{3k_B}\lambda \tag{9-9}$$

Where negative values of T_c are found, it is an indication that antiparallel alignment of magnetic moments, or antiferromagnetism, is present (see Equation 9-68). This means that the interactions of the spins or internal molecular fields are opposing each other and that the cooperative effect is absent. The Weiss constants for the antiferromagnetic case are discussed in Section 9.13.

The cooperative spin effects among the ions, which are responsible for the molecular field, are opposed by the thermal activity of the lattice (see Chapter 4, Volume I). As the temperature increases, the umklapp processes tend to increase and to detract from, and diminish, the degree of the cooperative spin effects. This disruptive behavior increases with increasing temperatures until the cooperative process effectively approaches zero. The temperature at which this takes place is the Curie temperature.

9.2. THE EFFECT OF TEMPERATURE UPON THE BRILLOUIN FUNCTION

The spontaneous magnetization for given temperatures was obtained (Section 9.1) by the simultaneous solution of Equations 8-112 and 9-5. An analysis of the behavior of the Brillouin function will be made with respect to its response to temperature.

Stated again for convenience, again using the notation $B_J(x)$ instead of $B_J(a)$,

$$M = Ng\mu_B J B_J(x) \tag{8-112}$$

where

$$B_J(x) = \frac{2J+1}{2J}\coth\left[\frac{2J+1}{2J}x\right] - \frac{1}{2J}\coth\frac{x}{2J} \tag{8-111}$$

For a given number of atoms per unit volume, N, where the maximum magnetic moment per atom is $g\mu_B J$, $B_J(x)$ must approach unity for M to be maximum. For this to be the case, x must approach infinity; and, for $x \to \infty$, it follows that $T \to 0$. The Brillouin function can be shown to approach unity in the following way, for very large x. From Equation 8-111

$$\coth\left[\frac{2J+1}{2J}x\right] \simeq \coth x \tag{9-10}$$

Equation 8-111 now may be approximated as

$$B_J(x) \simeq \frac{2J+1}{2J}\coth x - \frac{1}{2J}\coth x \qquad (x \to \infty) \tag{9-11}$$

Collecting terms

$$B_J(x) \simeq \frac{2J+1-1}{2J}\coth x = \coth x \tag{9-12}$$

It will be recalled that this may be reexpressed as

$$B_J(x) \simeq \coth x = \frac{e^x + e^{-x}}{e^x - e^{-x}} \tag{9-13}$$

For very large x, $e^{-x} \rightarrow 0$, so Equation 9-13 approaches unity. Thus, for very large x (Equation 9-4a) ($T \rightarrow 0$), the Brillouin function approaches its maximum value and the magnetization is maximum.

The upper limit of temperature now will be examined. For very large T

$$x = \frac{Jg\mu_B H}{k_B T} \rightarrow 0$$

Both hyperbolic functions in Equation 8-111 become quite small and

$$\coth\left[\frac{2J+1}{2J}x\right] \simeq \coth x \tag{9-14}$$

Then the Brillouin function again is approximated by

$$B_J(x) \simeq \frac{e^x + e^{-x}}{e^x - e^{-x}}$$

Applying L'Hopital's theorem, recalling that for large T, $x \rightarrow 0$

$$B_J(x) \rightarrow \frac{xe^x - xe^{-x}}{xe^x + xe^{-x}} = \frac{e^x - e^{-x}}{e^x + e^{-x}} = \frac{1-1}{1+1} = 0 \tag{9-15}$$

So, for high temperatures, $B_J(x) \rightarrow 0$ and the magnetization also approaches zero. This confirms the behavior derived from the graphical simultaneous solution of Equations 8-112 and 9-5.

The above analysis of the Brillouin function for $T \rightarrow 0$ implies that all of the moments are maximumly aligned at 0 K. As the temperature increases, the thermal excitation causes increasing amounts of antiparallel spin orientations. At, or above, the Curie temperature the number of parallel and antiparallel orientations can be considered to be equal to give zero spontaneous magnetization.

Thus, for maximum magnetization at 0 K,

$$M_o = Ng\mu_B J B_J(x) = Ng\mu_B J(1) \tag{9-16}$$

For any other temperature,

$$M_T = Ng\mu_B J B_J(x) \tag{9-17}$$

in which $B_J(x)$ is a function of temperature. The relative magnetization can be given by the quotient of Equations 9-17 and 9-16 as

$$\frac{M_T}{M_o} = B_J(x) \tag{9-18}$$

Thus, the Brillouin function gives the relative magnetization as a function of temperature.

The relative magnetization also can be expressed in another way by means of Equations 9-5 and 9-16

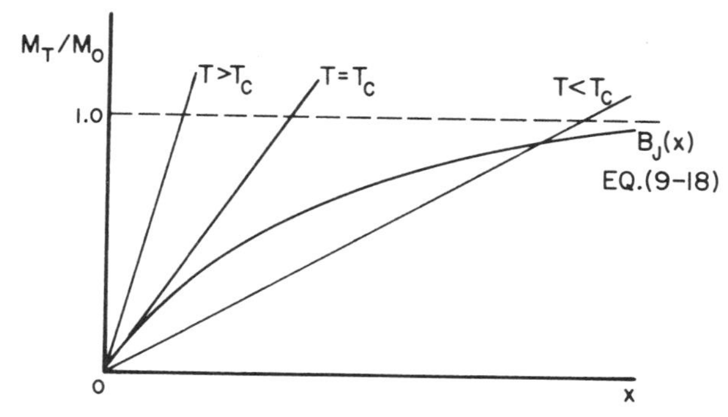

FIGURE 9-3. Graphical solutions of Equations 9-18 and 9-21. These provide the means for obtaining the theoretical data shown in Figure 9-2.

$$\frac{M_T}{M_o} = \frac{xk_BT}{Jg\mu_B\lambda} \div Ng\mu_BJ = \frac{xk_BT}{Ng^2\mu_B{}^2J^2\lambda} \tag{9-19}$$

Now, solving Equation 9-9 for λ,

$$\lambda = \frac{3k_BT_c}{Ng^2J(J+1)\mu_B{}^2} \tag{9-20}$$

and substituting Equation 9-20 into 9-19, it is found that

$$\frac{M_T}{M_o} = \frac{xk_BT}{Ng^2\mu_B{}^2J^2} \cdot \frac{Ng^2J(J+1)\mu_B{}^2}{3k_BT_c} = \frac{x(J+1)T}{3J\ T_c} \tag{9-21}$$

where x is given by Equation 9-4a, since for spontaneous magnetization, H = 0 (see Figure 9-2). This gives a comparatively simple expression for the relative magnetization as a function of temperature. The simultaneous solution of Equations 9-19 and 8-111 gives the relative magnetization as a function of T/T_c, as shown in Figure 9-3.

The case where $J \rightarrow \infty$ is, as previously noted (Figure 9-2), the case for a maximum number of moment orientations parallel to the field. It will be recalled (Section 8.3.4, Chapter 8) that the Brillouin function approaches the Langevin equation for this condition. The data for nickel and iron most closely agree with the curve for J = ½. This indicates that the origin of the ferromagnetic behavior must arise from spin alone and does not include a component from angular orbital momentum (see Sections 8.3.4, 8.6.2, and 8.6.4, Chapter 8). An electron in an inner band, such as an unscreened 3d or 4f level, has no effective orbital momentum since this is quenched by the crystal field (Section 8.6.2 in Chapter 8) at temperatures below the Curie temperature. As noted in the above sections on paramagnetism, such 3d and 4f levels appear to be inoperative with respect to orbital momentum; spin alone accounts for this magnetic behavior.

At the Curie temperature, and higher temperatures, the ferromagnetic behavior vanishes and paramagnetic behavior is observed. The paramagnetic susceptibility is given by Equation 8-102. The reciprocal of this equation is linear in T, and its temperature intercept, T_P, is expected at T_c;

$$\frac{1}{\chi_P} = \frac{3k_BT}{Ng^2J(J+1)\mu_B{}^2} \tag{9-22}$$

This is not exactly the case for iron, cobalt, nickel, and gadolinium. The actual

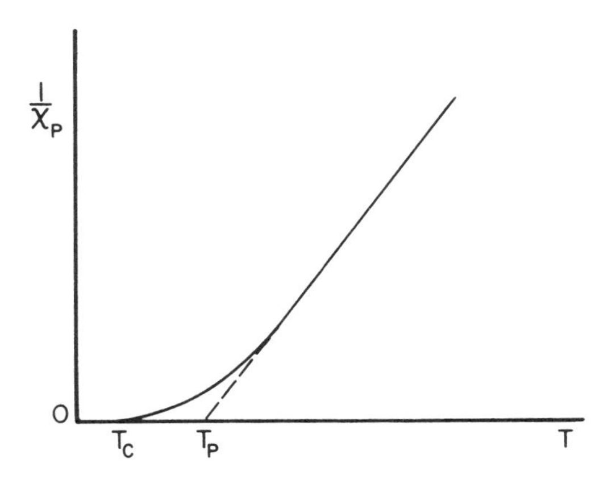

FIGURE 9-4. Reciprocal paramagnetic susceptibility of ferromagnetic elements as a function of temperature. (After Morrish, A. H., *The Physical Principles of Magnetism*, John Wiley & Sons, New York, 1965, 269. With permission.)

Table 9-1
CURIE TEMPERATURES AND EXTRAPOLATED VALUES FOR FERROMAGNETIC ELEMENTS

Element	T_c (K)	T_P (K)
Fe	1043	1093
Co	1403	1428
Ni	631	650
Gd	289	302.5

From Morrish, A. H., *The Physical Principles of Magnetism*, John Wiley & Sons, New York, 1965, 270. With permission.

behavior is shown schematically in Figure 9-4. The linear extrapolation of Equation 9-22 is used to determine T_P. Data on this behavior are given in Table 9-1.

This nonlinear behavior at the lower end of the temperature range of paramagnetism requires a correction in Equation 9-8. In this range

$$\chi_P \propto (T - T_c)^{-4/3} \tag{9-23}$$

instead of the −1 exponent in Equation 9-8.

The changes in the magnetic behavior of a ferromagnetic material are shown, over a wide range of temperature, in Figure 9-5.

9.3. EXCHANGE ENERGY

The examination of the exchange energy requires a model in which each band is treated as two half-bands. The electrons in each half-band have spins of the same direction and opposite to those in the other band (this is shown schematically in Figure

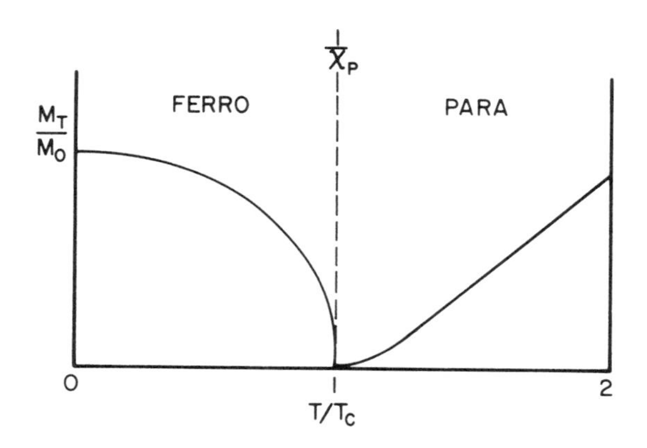

FIGURE 9-5. Schematic representation of the magnetic behavior of a ferromagnetic material over a wide range of temperatures.

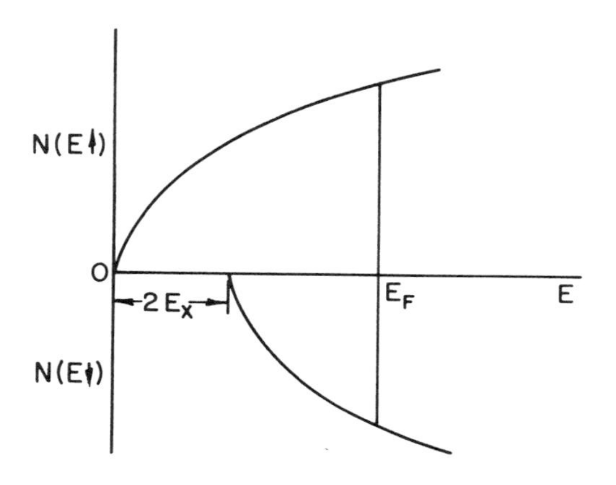

FIGURE 9-6. Density-of-states curves for the half-bands of opposite spin. The imbalance is responsible for ferromagnetism.

9-6). In the nonmagnetic state the initial number of spins of each kind is the same. An imbalance of spin gives rise to the ferromagnetic effect. As noted in previous sections, Fe, Co, and Ni contain incompleted 3d levels. The resulting spin imbalance arising from different numbers of electrons in each half-band is responsible for the ferromagnetic behavior of each element.

Coulombic forces are independent of spin and do not enter into this reaction. The exchange forces which occur result from the numbers of electrons with parallel and antiparallel spins. The antisymmetric wave functions demanded by the exclusion principle give rise to the exchange forces (Section 3.11, Chapter 3, Volume I). The result is that the probability of finding two electrons with parallel spins near to each other is small. Repulsive forces are set up and the exchange integral is negative (see Section 9.5).

The change in energy that occurs during the magnetization or demagnetization process is important because it affects the internal energy of the solid and those properties which reflect this. Assume that one half-band contains n more electrons than the other. In terms of the Weiss field

$$\lambda M = \lambda n \mu_B \tag{9-24}$$

where λ is the Weiss field constant and M is the magnetization. As more and more electron pairs are formed the energy, E_x, or exchange energy, decreases in magnitude and the ferromagnetism vanishes at the Curie temperature.

Starting with the Weiss model (Equation 9-2) in which the external field, H, equals zero

$$H_T = \lambda M \tag{9-25}$$

and summing the tendency for parallel-spin formation

$$E_x = -\int_o^M H_T dM = -\int_o^M \lambda M dM = -\tfrac{1}{2}\lambda M^2$$

$$= -\tfrac{1}{2}\lambda n^2 \mu_B^2 \tag{9-26}$$

The minus sign denotes that a decrease in energy, or parallel spin would not occur spontaneously.

If E_F is regarded as a constant, the transfer of electrons from spin up to spin down effectively lowers the energy of the top of the density of states of the spin-up half-band and raises that of the spin-down half-band. The change in the energy of each half-band is $\tfrac{1}{2}\lambda n^2 \mu_B^2$. The total difference in the energy between the bands, thus, is

$$2\,E_x = -\lambda n^2 \mu_B^2 = -\lambda M^2 \tag{9-26a}$$

as is shown in the figure.

This can be demonstrated in another way. Let N/2 be the initial number of spins in each half-band. If one half-band contains n more electrons than the other, then n/2 electrons will have flipped over into the other half-band. The half-band receiving these flipped electrons will contain N/2 + n/2 electrons. The half-band giving up these electrons will now have N/2 − n/2 electrons in it. The change in total energy will be

$$E_x = \frac{n}{2}\lambda \mu_B^2 \left\{ \left[\frac{N}{2} - \frac{n}{2}\right] - \left[\frac{N}{2} + \frac{n}{2}\right] \right\} \tag{9-27}$$

The quantity n/2 is used in the factor outside of the brackets to represent the number of electrons which have flipped from one half-band to the other. Then, collecting terms,

$$E_x = \frac{n}{2}\lambda \mu_B^2 \left[\frac{N}{2} - \frac{n}{2} - \frac{N}{2} - \frac{n}{2}\right] = \frac{n}{2}\lambda \mu_B^2 \left[-\frac{2n}{2}\right]$$

or, simplifying,

$$2E_x = -\lambda \mu_B^2 n^2 \tag{9-28}$$

which is the same as Equation 9-26a.

Now consider the transfer of an electron from one half-band to another of equal initial population. The excess number of electrons, n, now equals 2, since the first half-band lost an electron and the second gained one. Thus, from Equation 9-28

$$\Delta E_x = -\tfrac{1}{2}\lambda \mu_B^2 n^2 = -\tfrac{1}{2}\lambda \mu_B^2 \cdot 4$$

$$= -2\,\lambda \mu_B^2 \text{ (per electron)} \tag{9-29}$$

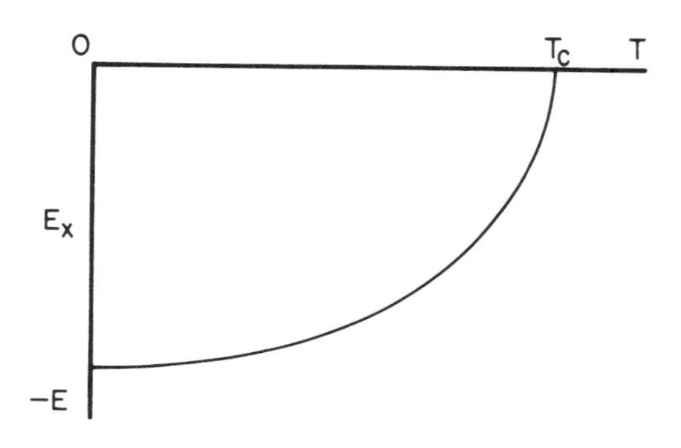

FIGURE 9-7. Exchange energy as a function of temperature.

The behavior of the exchange energy as a function of temperature can be obtained from combining Equations 9-26 and 8-112, or

$$E_X = -\tfrac{1}{2}\lambda M^2 = -\tfrac{1}{2}\lambda\left[Ng\mu_B J B_J(x)\right]^2 \qquad (9\text{-}30)$$

where

$$B_J(x) \rightarrow \left| \begin{array}{l} 1,\ T \rightarrow 0 \\ 0,\ T \rightarrow T_c \end{array} \right.$$

See Section 9.2. This relationship enables the graphical representation of the expression for E_x as a function of temperature as given in Figure 9-7.

In one case energy must have been added to the electrons to have effected the transfer noted by Equation 9-29. The only other case is that of spontaneous magnetization. Here the internal energy content must have been lowered by the transfer; the net energy only could have been lowered by the spontaneous transfer. Thus, from Equation 9-29

$$-\Delta E_X < 2\lambda\mu_B{}^2 \qquad (9\text{-}31)$$

Then, from the differential of Equation 3-26 with respect to $\bar{k}$, it may be approximated, in a simple way, that

$$-\Delta E_X = \frac{2h^2\bar{k}}{8\pi^2 m}\Delta\bar{k} < 2\lambda\mu_B{}^2 \qquad (9\text{-}32)$$

The transfer would have to be spontaneous because the energy of the system would have been lowered, and this is expected to be a function of the wave vector. The wave vector which gives the greatest lowering of the energy is preferred.

It follows, from Equation 9-32, that the exchange energy should vary with the crystallographic direction because of its dependence upon the wave vector. The energies required to form a domain in single crystals of various orientations, therefore, are expected to vary. These are the crystalline orientation energies. Examples of this anisotropy are shown in Figure 9-8.

It is apparent that certain crystalline directions are easier to magnetize than others. These are called easy and hard directions of magnetization. The differences in energy between hard and easy directions of magnetization are known as anisotropy energies. This effect is most apparent in the case of cobalt in Figure 9-8. Anisotropy energies

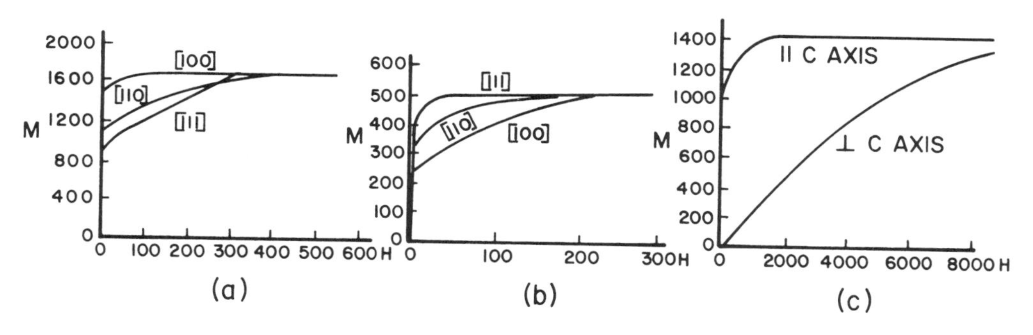

FIGURE 9-8. Magnetization curves of single crystals of (a) iron; (b) nickel; and (c) cobalt (M in G, H in Oe) The easiest directions for spontaneous magnetization are those with the largest values of M for H = 0. (After Honda, K. and Kaya, S., *Sci. Rep. Tôhoku Univers.*, 15, 721, 1926.)

are discussed in Section 9.7. An approximation of anisotropy energy is given by the second term in Equation 9-55.

9.4. HEAT CAPACITY INCREMENT

As previously noted, the addition of thermal energy offsets the lowering of energy due to spin alignment. As thermal energy is added, some of it is absorbed in randomizing the spins, increasing the internal energy of the material. The contribution to the heat capacity made by this reaction can be obtained from Equation 9-26 as follows:

$$C_M = \frac{\partial E_x}{\partial T} = \frac{\partial}{\partial T}\left[-\tfrac{1}{2}\lambda M^2\right] = -\lambda M \frac{dM}{dT} \quad (9\text{-}33)$$

The behavior of C_M as a function of temperature can be derived from Figure 9-7, where E_x is given as a function of temperature, because C_M is the derivative of E_x with respect to T. At T = 0 K, the slope is zero. The slope becomes increasingly positive as T increases and becomes very large at $T = T_c$. For temperatures at and beyond T_c, $E_x = 0$ and the slope is zero; the substance is paramagnetic beyond T_c. This variation of C_M with T is illustrated in Figure 9-9.

Such strong effects of magnetic changes upon the heat capacity of a ferromagnetic material also are manifested in other properties. Using the approximation $C_P \simeq C_V$ (Equation 4-7) both the coefficient of thermal expansion (Equation 4-135) and the thermal conductivity (Equation 4-151) are influenced by these changes, since both properties are functions of the heat capacity.

The experimental behavior shown in Figure 9-9 can be explained if it is assumed that neighboring parallel spins produce local molecular fields, rather than by using the Weiss assumption (Equation 9-1) and that the moments are parallel to these spins. The persistence of some of these spin clusters above T_c could account for the difference between the observed heat capacity and that derived from the Weiss theory. They also can explain the nonlinear thermoelectric properties of nickel and nickel-base alloys above the Curie temperature as discussed in Section 7.13.1 and shown in Figures 7-11 and 7-12, Chapter 7.

If the increment of heat capacity due to ferromagnetic effects is added to the lattice and electronic heat capacities (Chapters 4 and 5, Volume I), then the heat capacity of ferromagnetic materials may be obtained. This is shown in Figure 9-10. This effect is very pronounced in Fe, Ni, and Gd. This increment constitutes an appreciable percentage of the heat capacity of these solids. The extent of its influence is readily appreciated by recalling that $C_V \simeq C_P \simeq 3R$ for $T \geqslant \theta_D$.

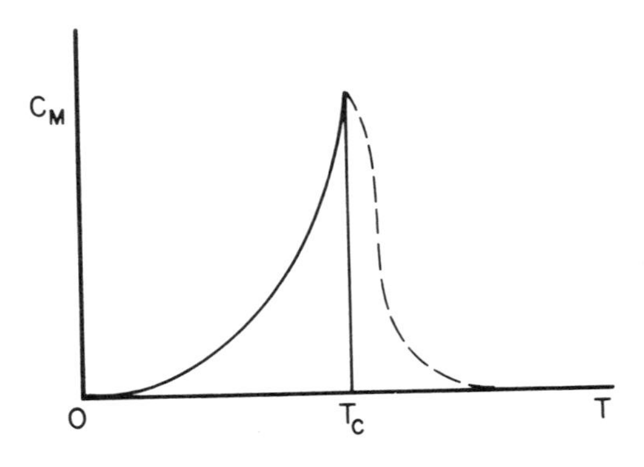

FIGURE 9-9. Weiss theory for the contribution to the heat capacity made by ferromagnetic effects. The broken curve shows experimental behavior above T_c.

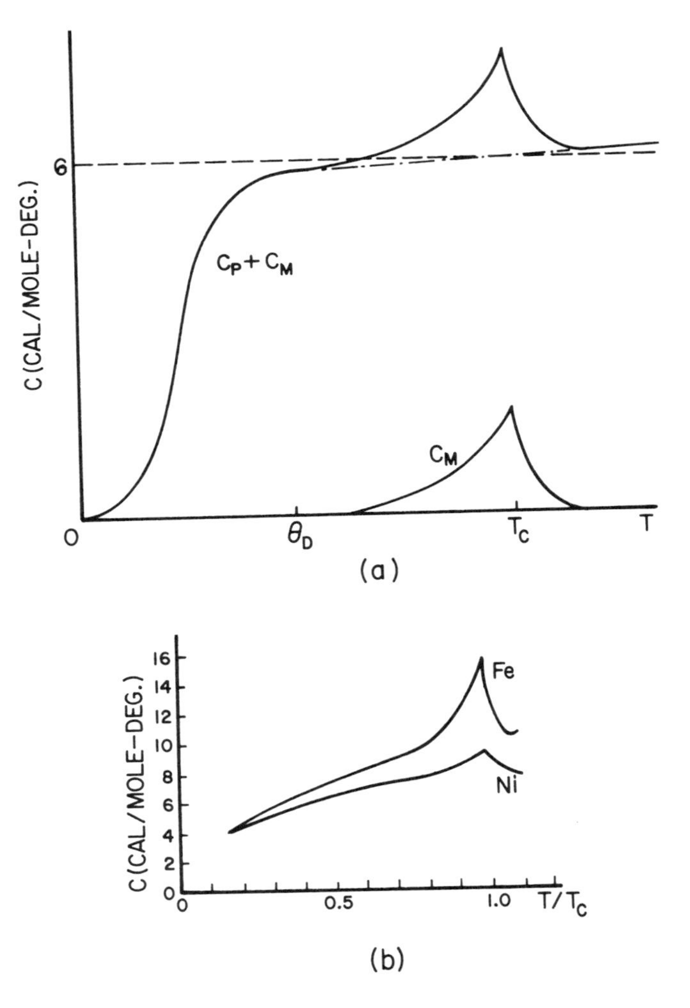

FIGURE 9-10. Heat capacity effects. (a) Schematic influence of ferromagnetic effects; (b) Heat capacities of iron and nickel. (After Morrish, A. H., *The Physical Principles of Magnetism*, John Wiley & Sons, New York, 1965, 272. With permission.)

9.5. THE EXCHANGE INTEGRAL

The spontaneous alignment of the magnetic moments within a domain, as postulated by Weiss, was explained by Heisenberg on an energy basis. The potential energy, E_J, of two electrons with spin S_i and S_j is

$$E_J = -2J_eS_iS_j\cos\phi_{ij} \tag{9-34}$$

Here, ϕ_{ij} is the angle between the spins and J_e is the exchange integral. This interaction is primarily coulombic. The two spins appear to be directly coupled, but the spins must be considered as required by the exclusion principle. When J_e is positive, the energy is minimized and the spins are parallel. The substance is ferromagnetic. When J_e is negative the stable state is for the spins to be antiparallel and the substance is antiferromagnetic. Diamagnetic behavior is present when J_e is approximately zero.

The Pauli Exclusion Principle allows a given state to be occupied by electrons of opposite spins; those with a given spin must occupy other states. If these states are considered as 'orbits' , the average distance between electrons will be different for the parallel and antiparallel cases. The coulombic energy will be different for both cases because of the differing distances. The exchange integral may be considered to be a measure of the way in which the electrostatic energy is affected by spin orientation. The minimization of this spin energy accounts for the domains. For n nearest neighbors, neglecting all others, and considering only the components of spin parallel to the magnetization,

$$E_J = -2nJ_eS^2 \tag{9-35}$$

where the spin momentum is measured in units of $h/2\pi$. The magnetic moment for one spin is

$$M = Sg\mu_B \tag{9-36}$$

Recalling that the product of H and M is energy, it follows that

$$H_TM = H_TSg\mu_B = k_BT_c = E(T_c) \tag{9-37}$$

at the Curie temperature. By means of Equations 9-35 and 9-37

$$E_J = -2nJ_eS^2 = -H_TSg\mu_B \tag{9-38}$$

For spontaneous magnetization (Equation 9-2) $H_T = \lambda M$. In order to relate this to the given n ions, this must be in terms of the atomic volume, V. Thus,

$$H_T = \lambda\frac{M}{V} = \lambda\frac{Sg\mu_B}{V} \tag{9-39}$$

and, when Equation 9-39 is substituted into Equation 9-38,

$$E_J = -2nJ_eS^2 = -\lambda\frac{Sg\mu_B}{V}Sg\mu_B \tag{9-40}$$

Or,

$$2nJ_eS^2 = \frac{\lambda S^2g^2\mu_B^2}{V} = \lambda S^2g^2\mu_B^2n \tag{9-41}$$

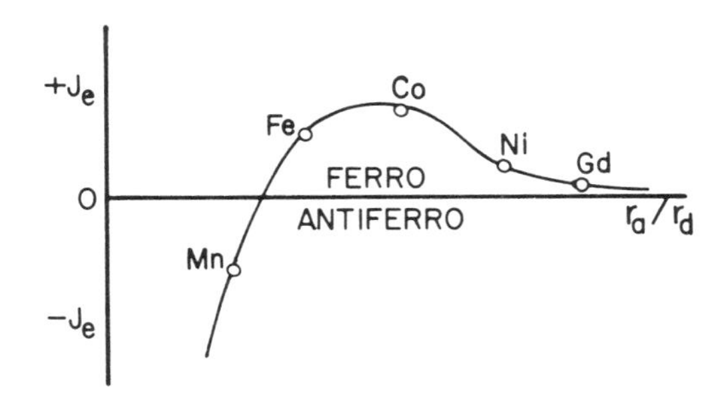

FIGURE 9-11. The exchange integral as a function of the ratio of half the internuclear distance to the d-shell radium according to Bethe.

since 1/V = n. Rewriting Equation 9-9, and recalling that ferromagnetic behavior arises from spin alone and not angular orbital momentum, S is used instead of J to give

$$\lambda = \frac{3k_B T_c}{Ng^2 J(J+1)\mu_B^2} \equiv \frac{3k_B T_c}{Ng^2 S(S+1)\mu_B^2} \qquad (9\text{-}42)$$

and substituting this into Equation 9-41, and simplifying,

$$J_e = \frac{3k_B T_c}{Ng^2 S(S+1)\mu_B^2} \cdot \frac{g^2 \mu_B^2 n}{2n} = \frac{3k_B T_c}{2NS(S+1)} \qquad (9\text{-}43)$$

This provides a relationship between the exchange integral and the Curie temperature for the molecular field theory. For the most elementary case, that of the simple cubic lattice, n = 6. For S = ½, $J_e/k_B T_c$ = 1/3. Other, more exact, calculations give this ratio as slightly larger than 1/2.

For most materials J_e is negative; the spins are antiparallel and the substance is antiferromagnetic. Where J_e is positive ferromagnetic behavior is probable.

Bethe showed that J_e should be positive when half the internuclear distance, r_a, is large with respect to the radius of the d electron orbits, r_d, giving rise to the exchange effects. Here r_d is the average radius of the d shell. In other words, the spin conditions responsible for ferromagnetic behavior occur when the spatial conditions are correct for the minimization of d-level overlap. Slater (1930) calculated that the ratio of r_a/r_d had to be at least equal to 3.0, but not much greater. His data are:

Metal	Fe	Co	Ni	Cr	Mn	Gd
r_a/r_d	3.26	3.64	3.94	2.60	2.94	3.1

Figure 9-11 shows that only Fe, Co, Ni, and Gd are in the region of positive values of J_e. The ratio for Gd does not agree with that of Slater. The ratio noted above implies that the d-levels are narrow with respect to the internuclear distance and that those do not overlap or interact between nearest neighbors. The other transition elements do not satisfy these conditions and are not ferromagnetic.

The relationship between J_e and r_a/r_d is of importance in the explanation of the ferromagnetic properties of alloys of nonferromagnetic transition elements. Suitable alloying changes the ratio and can result in a positive value for J_e. This affects λ (Equation 9-41) and results in spontaneous magnetization (Equation 9-1). See Section 9.10 in which chromium-base and manganese-base alloys showing ferromagnetic properties are discussed.

Bloch (1930) devised another means for describing the magnetization as a function of J_e. This is known as the spin-wave approach and is valid only at very low temperatures. While this work will not be discussed in detail, the concept of spin waves is of interest. They can affect thermal conductivity at very low temperatures.

For the purposes of this discussion it is assumed that all of the spins of a ferromagnetic material are ordered at 0 K. Now, assume that a small increase in temperature results in the reversal of just one spin. Every ion in the solid has the same probability of having the spin of one of its electrons reversed. This leads to the idea that the electron with the reversed spin will not remain associated with its original ion. If it did, the exchange forces would tend to return it to its original spin direction. The original small increase in temperature prevents this and the electron passes from one neighboring ion to the next. The motion of the electron through the lattice causes the spin vectors of the other electrons to precess such that a line joining all the precessing vectors at a given instant will have a shape like a wave. These are called spin waves.

These spin waves are limited to certain wavelengths because of the boundary conditions in a manner similar to lattice vibrations. Quantized spin waves are known as "magnons", and, as would be expected, behave in a manner corresponding to phonons. The possibility of the pairing of two reversed spins on adjacent ions is neglected. The magnons are completely degenerate because their total number is not fixed. They obey the Bose-Einstein statistics since the interchange of two opposite spins does not change the state of the system; this method is used to calculate their density of states and their probable number.

Each magnon changes the spontaneous magnetization. The relative change in the spontaneous magnetization, given by the Bloch $T^{3/2}$ law for BCC and FCC latices, is

$$\frac{\Delta M}{M_o} = \frac{0.1174}{n_c}\left[\frac{k_B T}{J_e}\right]^{3/2}$$

where M_o is the magnetization at 0 K and n_c is the number of ions per unit cell. Magnons also contribute to the thermal conductivity at low temperatures, if the relaxation times are short, as in certain garnets and antiferromagnetic materials.

9.6. DOMAINS

As previously noted, the existence of the spontaneously magnetized volumes, or domains, has been verified by the use of Bitter patterns. The exchange integral, the exchange energy, and anisotropy energy all favor the formation of domains. Each of the domains in a material is spontaneously magnetized and its magnetic effect may be summarized by a vector. The sum of the vectors of all of the domains gives the magnetization of the specimen.

The magnetization process can occur by the movement of the walls of those domains with vectors parallel to the applied field or by the rotation of the domains. As the domain walls move, adjacent domains are incorporated into the one whose walls are moving. These effects are shown in Figure 9-12.

The unmagnetized specimen shows the magnetization vectors cancelling each other. In the second illustration, the domain with the magnetization vector parallel to the applied field has enlarged at the expense of the other domains. The motion of the domain walls is not continuous, but occurs in small, discrete steps. These occur as a result of irreversible wall motions, or of irreversible rotations within a domain. This discrete behavior is known as the Barkhausen effect. The third sketch shows that the direction of magnetization of a domain will rotate to a hard direction in order to become parallel to the applied field (see Figure 9-14).

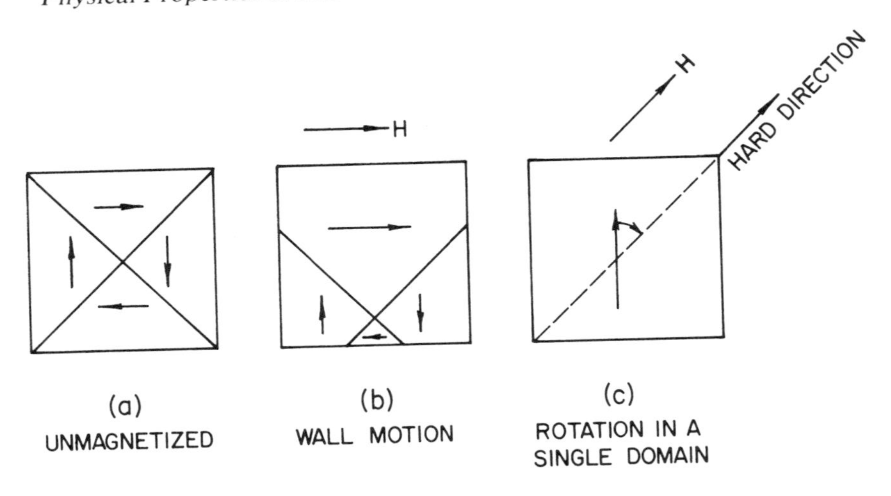

FIGURE 9-12. Domain motion. (Dekker, A. J., *Solid State Physics*, Prentice-Hall, Englewood Cliffs, N.J., 1957, 476. With permission.)

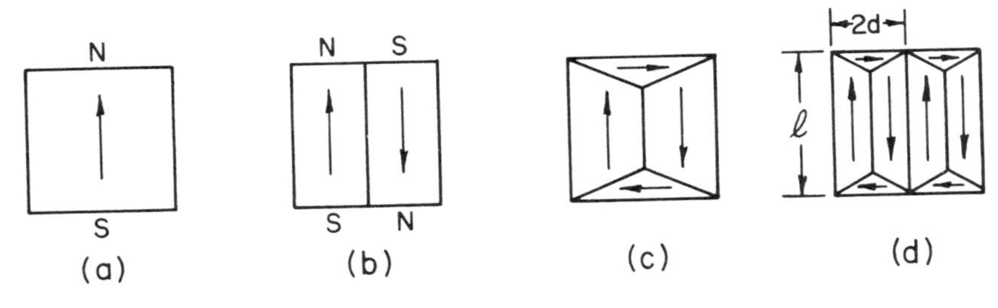

FIGURE 9-13. Domain structures showing progressively lower energy configurations.

The existence of domains also has been explained on the basis of minimized free energy. This is done by minimizing the internal magnetic energy of the solid. The entropy term is neglected because for $T < T_c$, $M \rightarrow M_o$, and a high degree of order exists.

Figure 9-13a shows a crystal which is a single domain. It has a high value of magnetic energy because of the large external field. When this is divided into two domains, the magnetic energy becomes about half of the single domain. This process can be continued so that for n domains the energy becomes approximately 1/n of that of the single original domain. The limit to this energy reduction is reached when the energy required for the creation of an additional domain boundary is greater than the lowering of the internal energy. The demagnetization energy also is reduced as the number of domains is increased; smaller domains of closure are involved in demagnetization.

Where domains of closure exist (such as in Figure 9-13d) the magnetic energy is zero. The magnetic components normal to the triangular boundaries are such that no poles are formed. The magnetic circuit lies entirely within the specimen and the specimen is demagnetized.

The domains tend to line up in directions of easy magnetization, the long directions in Figure 9-13. The energy required to form domains of closure depends upon the crystal structure of the specimen. This, in turn, is determined by the anisotropy energy, since these must represent harder directions of magnetization (see Section 9.7).

9.7. DOMAIN BOUNDARIES AND SIZES

The interface, or domain wall, between adjacent domains is known as the Bloch wall. The spin orientations on either side of the boundary are different. This is not a sudden, discrete, change from one spin orientation to the other, but a gradual transition of spins across the boundary.

The structure of the Bloch wall represents a series of spin orientations, each of which is slightly different from the next to permit a gradual transition from the spin orientation of one domain to that of its neighbor. Thus, each spin in the boundary has a slightly different orientation than the one next to it. Based upon Equation 9-34 this may be expressed as

$$E_x = -2J_eS^2 \cos \phi_{ij} \tag{9-44}$$

in which ϕ_{ij} is the angle between the spin magnetic moments of adjacent ions. The lowest energy condition for positive values of J_e is when ϕ_{ij} is zero.

Using the series approximation

$$\cos \phi_{ij} = 1 - \frac{\phi^2_{ij}}{2} \ldots$$

Equation 9-44 becomes, considering only neighboring spins,

$$E_x \simeq J_eS^2\phi^2_{ij} + \text{constant} \tag{9-45}$$

If it is assumed that the change in spin orientation across the domain wall occurs in equal increments over N ions, then

$$\phi_{ij} = \frac{\phi}{N} \tag{9-46}$$

And, the energy between adjacent pairs of ions is

$$E_x(1) \simeq J_eS^2 \frac{\phi^2}{N^2} \tag{9-47}$$

Or, across a domain boundary consisting of N pairs of ions, the energy is

$$E_x(N) = NJ_eS^2 \frac{\phi^2}{N^2} = J_eS^2 \frac{\phi^2}{N} \tag{9-48}$$

The unit energy stored in each one of the N planes in the boundary per unit area is

$$E_x(N,a) = \frac{J_eS^2\phi^2}{a^2N} \tag{9-49}$$

where a is the lattice constant.

The anisotropy energy will increase as the magnetic moments rotate out of easy directions of magnetization. Let the anisotropy energy be K per unit volume. The volume per unit area of wall for N ions, having an interionic distance a, in a simple cubic lattice is

$$\frac{Na^3}{a^2} = Na \tag{9-50}$$

so that the anisotropy energy per unit area is

$$E_{anis} = KNa \tag{9-51}$$

This energy component tends to minimize the domain wall thickness. The total energy per unit area stored in the domain boundary then is the sum of Equations 9-49 and 9-51:

$$E_T = \frac{J_e S^2 \phi^2}{a^2 N} + KNa \tag{9-52}$$

The extent of the Bloch wall can be determined by minimizing the stored energy:

$$\frac{\partial E_T}{\partial N} = -\frac{J_e S^2 \phi^2}{a^2 N^2} + Ka = 0 \tag{9-53}$$

$$\frac{J_e S^2 \phi^2}{a^2 N^2} = Ka$$

$$N^2 a^2 = \frac{J_e S^2 \phi^2}{Ka}$$

$$Na = \left[\frac{J_e S^2 \phi^2}{Ka}\right]^{1/2} \tag{9-54}$$

which gives the domain wall thickness. In iron the domain wall is about 200 to 300 ion spacings and has a total stored energy, E_T, of about 2 to 4 erg/cm^2.

Another more explicit expression can be obtained for E_T by the use of Equation 9-54 in Equation 9-52:

$$E_T = \frac{J_e S^2 \phi^2}{a^2} \cdot a \left[\frac{Ka}{J_e S^2 \phi^2}\right]^{1/2} + \frac{1}{a}\left[\frac{J_e S^2 \phi^2}{Ka}\right]^{1/2} Ka \tag{9-55}$$

or

$$E_T = \frac{1}{a}\left[KaJ_e S^2 \phi^2\right]^{1/2} + \frac{1}{a}\left[J_e S^2 \phi^2\, Ka\right]^{1/2}$$

which reduces to

$$E_T = 2\left[\frac{KJ_e S^2 \phi^2}{a}\right]^{1/2} = 2S\phi\left[\frac{KJ_e}{a}\right]^{1/2} \tag{9-56}$$

The above approximation for the energy per unit area of domain wall assumed a simple cubic lattice and an equal spin increment. Corrections can be made for BCC and FCC lattices by using multiplication correction factors of 2 and 4, respectively. The assumption of equal increments of spin orientations across the wall is only approximate because of the crystalline anisotropy. The interionic distance will vary with crystalline direction and affect E_T.

The motion of domain walls discussed in the previous section now may be visualized on the basis of small changes in the spin directions across the domain wall. This explains the relatively low energy required to move the wall and to absorb a part of an adjacent domain. Such small increments require considerably less energy for motion into the next domain than that required for the simultaneous rotation of all of the magnetic moments in a domain (see Section 9.8).

It was noted in the previous section that increasing the number of domains of closure decreases the internal magnetic energy of a solid. However, as the number of such domains increases, the quantity of domain wall "material" must, therefore, increase accordingly. In terms of Figure 9-13d, the domain wall energy will, therefore, vary inversely with d and directly with ℓ.

Consider the domains of closure apparent upon the polished surface of a ferromagnetic single crystal. An approximation for the energy of the wall, E_W, per unit area of domain surface may be given by

$$E_W = E_T \ell / d$$

Since the long dimension of the domain will lie in the easier directions of magnetization, the domains of closure will contain the harder directions of magnetization; thus, $\ell \gg d$. Because d is so small, and the anisotropy energy decreases as d decreases the domain of closure may be approximated as being a small oblong. Further, since each domain wall is shared by two adjacent domains, the wall width associated with a given domain is d/2. Using the anisotropy energy, K, the energy of a wall of a domain of closure may be estimated by

$$E_c = Kd/2$$

The energy of a domain boundary, per unit area of domain surface, is given by the sum

$$E_D = E_W + E_c = E_T \ell / d + Kd/2$$

This is minimized with respect to the domain width, for the reasons noted above, to give

$$\frac{\partial E_D}{\partial d} = -E_T \ell / d^2 + K/2 = 0$$

and the minimum value of d is found to be

$$d = \left[\frac{2E_T \ell}{K}\right]^{1/2}$$

Using iron for purposes of illustration, $E_T \simeq 2$ ergs/cm², $K \simeq 4 \times 10^5$ ergs/cm³, and assuming $\ell = 1$ cm, $d \simeq 0.003$ cm.

The substitution of d back into the expression for E_D results in

$$E_D = E_T \ell \left[\frac{K}{2E_T \ell}\right]^{1/2} + \frac{K}{2}\left[\frac{2E_T \ell}{K}\right]^{1/2} = \left[\frac{E_T \ell K}{2}\right]^{1/2} + \left[\frac{E_T \ell K}{2}\right]^{1/2}$$

$$= (2E_T \ell K)^{1/2}$$

For the same data used above, $E_D \simeq 13 \times 10^2$ ergs/cm².

Such uniform domains are not always present in single crystals. Much depends upon the orientation; many other shapes may be obtained. In polycrystalline materials each grain will contain domains along its easy direction of magnetization and free poles will appear when the orientations of the adjacent grains do not provide closure. When the number of free poles at the surface of a material is small compared to those at grain boundaries, the grain size determines domain configuration. In sheet, or foil, which is thinner than the average grain size, the thickness determines the domain structure.

Domain shapes in polycrystalline materials also are affected by such factors as chemical homogeneity, precipitates, inclusions, voids, and internal stresses (local anisotropy). The domain structure in multiphase alloys depends upon the amount, size, and distribution of the ferromagnetic phase, in addition to the other factors. In each case, the domain structure is that which tends to minimize the total energy.

Thin garnet films (Sections 9.12 and 9.14.2.1) of the order of 10^{-4} cm thick have domains very unlike those shown in Figure 9-13d. These assume very thin, convoluted configurations in which domains of opposite directions of magnetization (perpendicular to the plane of the film) are adjacent to each other. These are visible when observed by means of a microscope when polarized light is used. They provide practical examples of applications of domain theory.

Consider such a film in which the easy direction of magnetization is perpendicular to its surface. The application of an external magnetic field perpendicular to the film (parallel to the easy magnetization direction) will cause the favorably oriented domains to absorb those which are unfavorably oriented. As a result, the unfavorably oriented domains become thinner. As the applied field is increased, the thinning continues, but in a nonuniform way. Some portions of the unfavorably oriented domains become thinner more rapidly than other portions; eventually these domains split into wormlike segments. These, in turn, continue to become absorbed by the favorably oriented material and contract symmetrically. If the external field is removed before the absorption is complete, small, cylindrical domains, whose magnetization direction is opposite to that of the matrix, will remain. These appear as very small circles when viewed optically and are called bubbles.

Bubble domains are of the same nature as those in most ferromagnetic materials. The Bloch wall plays an important role here because the bubble is a small single domain within a matrix which is a large single domain of opposite magnetization. This explains the high degree of mobility of bubbles in contrast to the domains observed in other types of magnetic materials. Thus, they may be moved across the film, in two dimensions, by the application of small amounts of energy.

The stability of the bubble domains results from an equilibrium between the exchange and anisotropy energies. Large values of E_x (Equation 9-48) increase E_T (Equation 9-52), and increase the value of d. Increasing K increases E_{anis}. The cylindrical shape arises from the minimization of the energy of the domain wall.

The bubbles tend to maintain a minimum distance of approach because all have the same direction of magnetization; they will repel one another when they come too close together. Use is made of this repulsion in bubble circuitry. Also, when many bubbles are present in a thin-film crystal, they tend to array themselves so that their number per unit area is nearly constant. The application of a field parallel to the plane of the film causes the bubbles to move across the film parallel to the field.

The most effective garnets for this application are of the form $M_3Fe_5O_{12}$, where M is a rare-earth element (Section 9.12). Others are formed where M may be Bi or Bi in conjunction with a rare-earth element. These garnets can be made to contain bubbles of about 3×10^{-4} cm in diameter with densities of more than $10^5/cm^2$. Bubble velocities can be achieved which are at rates equal to or greater than that obtained from switching transitors, using less than 5% of the energy required by the transistors.

One method for using bubbles consists of grids of conducting loops which are deposited upon the garnet films. The loops are approximately the same size as the bubbles. Current is switched sequentially from one loop to an adjacent loop so that the bubble is attracted from the first to the second loop. In the binary notation used in most computers, a bubble on a given loop denotes "1" and its absence denotes "0". The circuitry involved in using such grids is very cumbersome. This technique has been called the conductor-access method.

Another method involves the deposition of small, separated, unsymmetrical, magnetic shapes (usually of Permalloy) such as Ys, bars, Ts, etc. to form bubble paths. A periodically rotating external field is applied parallel to the film. As the external field changes its direction, various portions of adjacent unsymmetric path components have opposite polarities. When this polarization is favorable, the bubbles will move to the next path component, and so on. The periodic magnetic field acts like a traveling wave and the bubble will remain at a given path component until the wave reaches its maximum. It then moves to the next path component. Bubble velocities have been achieved which are the equivalent of read-out rates greater than 10^5 bits/sec. This technique is called the field-access method.

Additional bubbles may be generated when using either of these methods. This is accomplished by splitting the bubbles in the field-access method. Here U-shaped conductors, with widths about equal to that of a bubble diameter, are used. The passage of a current through this configuration, when a bubble is present, causes the bubble to split into two. Energy from the applied field is absorbed by both bubbles so that their size is at least the minimum required for stability. Their mutual repulsion causes them to move in opposite directions.

A domain source is used for bubble generation in the field-access method. One such source consists of a circular Permalloy disc with a small projection extending from it. The diameter of the disc is about five times the diameter of a bubble. The domain beneath the disc is retained because a minimum magnetization always exists around the edge of the disc. As the external field rotates, the shape of the domain becomes distorted and part of it extends beyond the disc. When the magnetization is correct, a portion of the distorted domain becomes associated with an adjacent path component. Then, as the direction of magnetization continues to change, the distorted domain breaks in two to form a domain beneath the disc and a bubble at the adjacent path component. The process is repeated; the rate of bubble generation is a function of the frequency of the rotation of the external field.

The bubbles may be made to collapse in either of the above techniques. This is done by the application of a magnetic field of suitable strength so that the bubble size becomes smaller than that required for stability by the energy equilibrium noted above.

Without attempting to describe the technology required for applications of magnetic bubbles, based upon the properties described above, it can be seen that their application in such devices as computers may be expected to grow in importance.

9.8. MAGNETIZATION CURVES

Further evidence for the domain theory postulated by Weiss lies in the behavior of ferromagnetic materials under the influence of applied fields. Such materials can be made to show increasing amounts of magnetization, up to saturation magnetization, by the application of relatively weak fields. As noted in the preceding sections, this can take place by the absorption of unfavorably oriented domains by those which have favorable orientations with respect to the applied field. This occurs by motion of the domain, or Bloch, wall described in Section 9.7. The other mechanism occurs by the rotation of a domain to maximize its magnetic moment. This requires much greater energy than domain motion. The overall behavior is shown schematically in Figures 9-12 and 9-14.

At low applied fields the boundaries are moved nearly reversibly and the motion of the boundary movement can be varied 180°. The nature of the Bloch wall explains the relative ease of motion. As the external field is increased, a large increase occurs in M. This results from irreversible motion of the domain walls to new, stable positions. The domains grow in size. Further increases in the field cause diminishing increases in M, largely as a result of rotation, until saturation magnetization is reached. Here, further increases in the applied field have no further effect upon M and the material is said to be saturated.

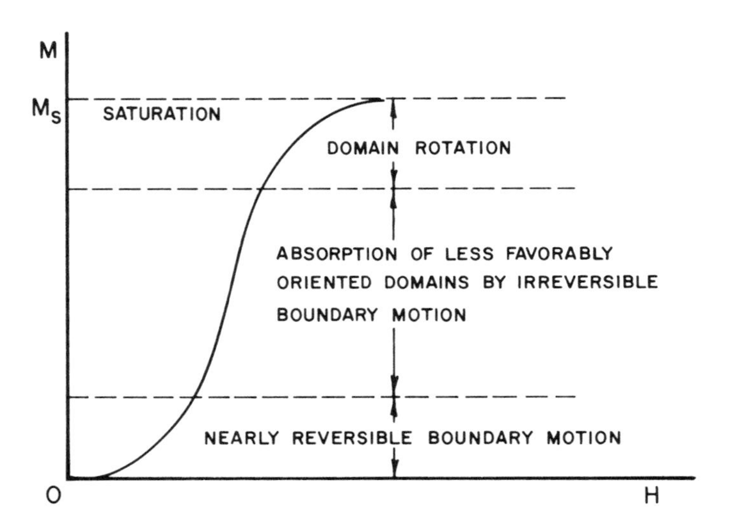

FIGURE 9-14. Mechanisms involved during magnetization when a field is applied to a demagnetized specimen. (Modified from C. Kittel, *Introduction to Solid State Physics*, 3rd ed., p. 489, Wiley, 1966.)

Before further discussion of magnetization curves, the units will be reviewed because the cgs units find wide use in engineering:

- B: magnetic flux density; 1 weber/m^2 = 10^4 G (cgs)
- H: magnetic field; 1 amp/m = $4\pi \times 10^{-3}$ Oe (cgs)
- M: magnetization; 1 weber/m^2 = $1/4\pi \times 10^4$ G (cgs)
- μ_o: permeability of vacuum: $4\pi \times 10^{-7}$ H/m = 1 G/Oe (cgs)

These units are related by $B = H + 4\pi M = \mu H$ (Equation 8-10) and $\mu = \mu_r\mu_o$ (Equation 8-14).

Plots of B vs. H (Figure 9-15) commonly are used in engineering to describe the properties of magnetic materials. Starting with a demagnetized material, the magnetic flux density follows the curve OB_s as H is increased until saturation is reached at B_s. This involves the mechanisms shown in Figure 9-14 and described above. When H is removed, the remaining magnetic flux is called remanence, B_r. This results from the irreversible boundary displacements. The reverse field required to demagnetize the material is called the coercive force, H_c. The continued application of the reverse field causes the material to become saturated in the opposite direction. The removal of the reverse field leaves a negative remanence. The reapplication of a field in the original direction will saturate the material again, if sufficiently strong. Magnetization curves are symmetrical about the origin and are called hysteresis loops. The work expended in going through such a cycle is expressed by

$$W = \oint H dM$$

where, in engineering measurements, M actually is that component of the magnetization parallel to the applied field, H. The energy absorbed in performing this work largely is dissipated as heat.

It can be seen from this that transformer materials with the most efficient properties should have minimum area hysteresis loops in order to minimize energy losses. Materials of this type are known as soft magnetic materials. These are designed to have small hysteresis effects and low H_c. Special Fe-Si and other alloys are available so that

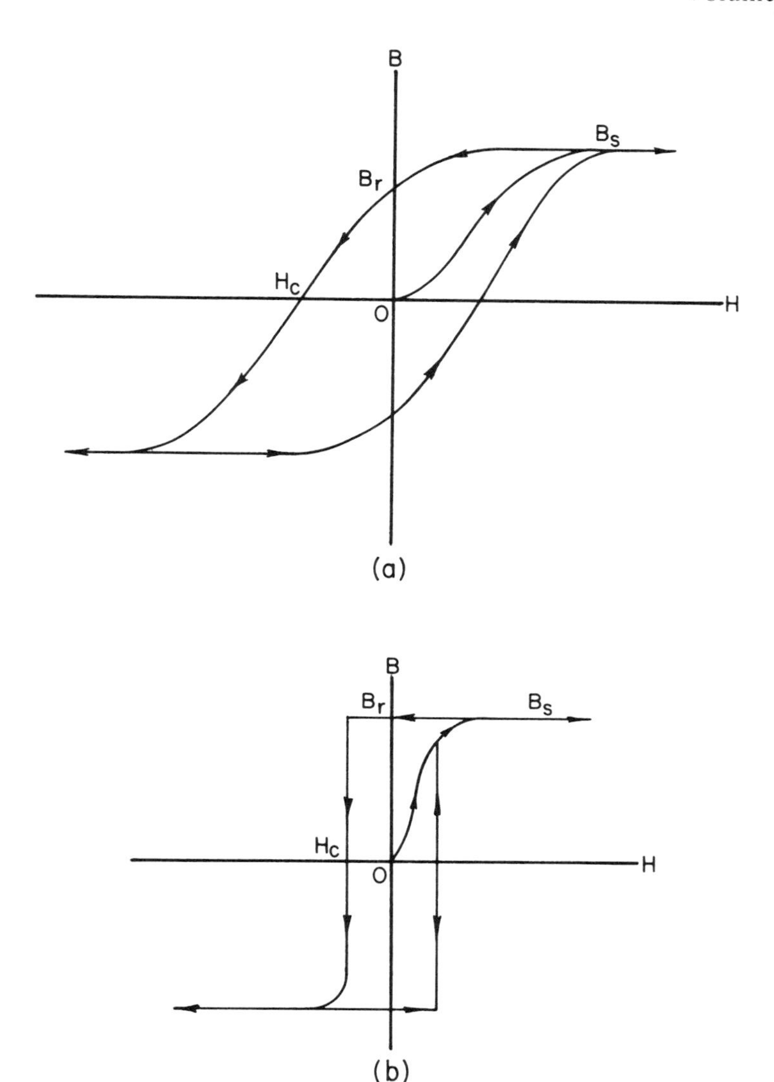

FIGURE 9-15. Hysteresis loops. (a) Normal; (b) rectangular. (Note: B scale is compressed.)

their easy directions of magnetization are in the direction of the sheet. The permeabilities of these materials is high (Equation 8-13). (See Section 9.14.2.)

Hard magnetic materials have large hysteresis loops with large B_r and H_c. These are used for permanent magnets. The permeability of such materials is low (see Section 9.14.1).

9.9. THE ELEMENTS Fe, Co, AND Ni

The elements of the first transition series have 4s bands which can accept 2 electrons and 3d levels which can accommodate up to 10 electrons. The hybridized nature of these bands is discussed in Chapter 7. Above the Curie temperature the numbers of electrons in each of the half-bands are equal. Below this temperature one of the d-level half-bands contains more electrons than the other, leaving an imbalance in the spin. The occupation of the bands has been deduced from observed magnetic and other properties. The net spin imbalance is given in Table 9-2.

Table 9-2
MAGNETIC MOMENTS OF IONS

Element	μ_B/ion
Fe	2.22
Co	1.71
Ni	0.61

Table 9-3
ELECTRONIC CONFIGURATIONS OF Fe, Co, AND Ni

	Ground states		Magnetic state						
Element	3d	4s	3d↑	3d↓	4s↑	4s↓	↑ 3d holes	3d holes ↓	Net no. holes
Fe	6	2	4.8	2.6	0.3	0.3	0.2	2.4	2.2
Co	7	2	5.0	3.3	0.35	0.35	0	1.7	1.7
Ni	8	2	5.0	4.4	0.3	0.3	0	0.6	0.6

After Bozorth, R. M., *Ferromagnetism*, Van Nostrand, New York 1951, 438. With permission.

The numbers of Bohr magnetons per ion represent the number of unbalanced 3d spins, or holes, in the crystalline states of these elements. The nonintegral values are averages resulting from the s-d hybridization. A simplified rationalization of this behavior is given in Table 9-3. The net number of holes, or unbalanced spins, equals the magnetic moments of the atoms given in Table 9-2 (see Figure 9-16). These configurations appear to be verified by thermoelectric and other nonmagnetic measurements.

Both magnetic and thermoelectric data suggest that about 0.6 3d holes exist in the Ni atom. This could be pictured as a mixture of atoms of $3d^9$ and $3d^{10}$ configurations in a suitable ratio. More probably the configuration of Ni arises from an assembly of mixtures of $3d^8$, $3d^9$, and $3d^{10}$ configurations. In addition, the electron configuration of a given atom may not be fixed (itinerant states). It is not difficult to picture electron interchanges, and their corresponding equivalent, hole hopping, resulting from interchanges between ions. This could arise from overlaps of the hybridized wave functions.

Cobalt, with a net number of holes of about 1.7 per atom could, similarly, be expected to show an assembly of combinations of $3d^8$ and $3d^9$ configurations.

In the case of Fe, it appears that the combination of electronic configurations should consist mainly of $3d^8$ configurations since the net number of holes is 2.2 per atom. A combination of $3d^7$ and $3d^9$ states also has been suggested to explain this. This could be the more probable because the interaction of the hybridized 4s levels tends to minimize the energy of the $3d^7$ - $3d^9$ combinations.

The simplified models of the band structures of these elements given in Figure 9-16 are configurations which are derived from an averaging of the various electron configurations. They provide models consistent with the physical properties of these elements and their alloys.

9.10. ALLOYS OF TRANSITION ELEMENTS

The magnetic properties of binary alloys of Fe, Co, and Ni are a direct function of the amount of the alloying element present. This is shown for nickel-base alloys in Figure 9-17.

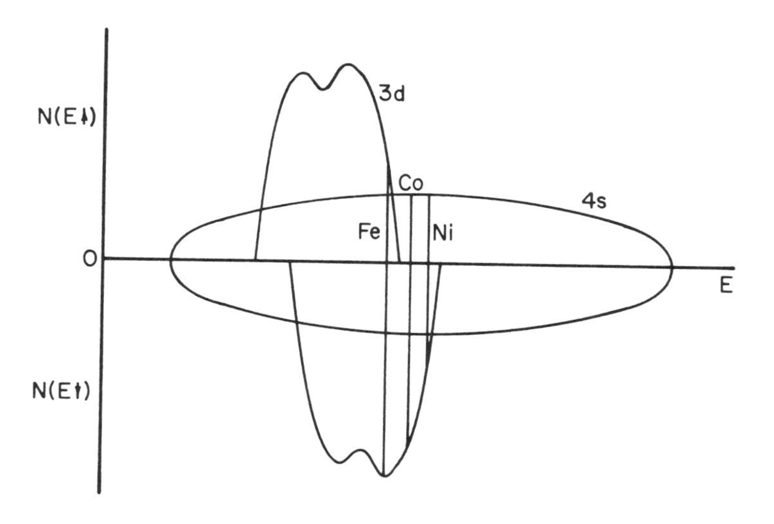

FIGURE 9-16. Schematic diagram of the band structures of Fe, Co, and Ni.

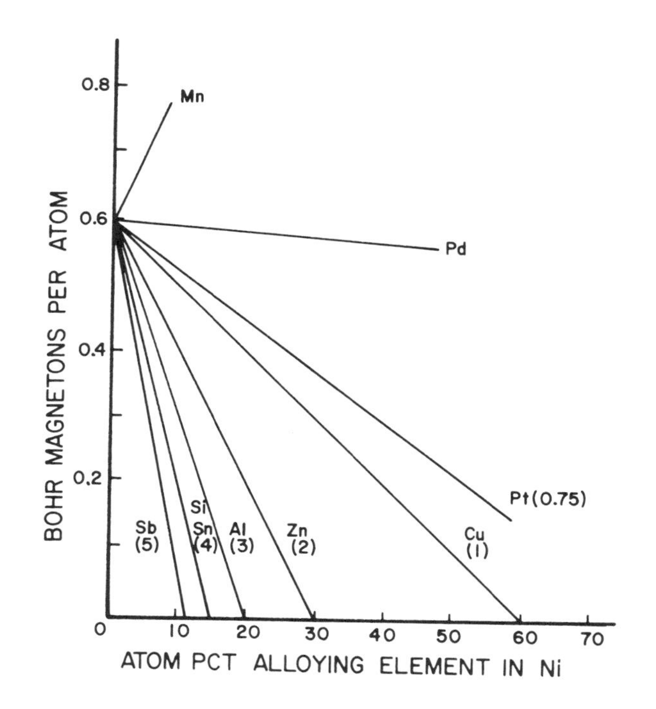

FIGURE 9-17. Effect of binary alloy additions on the magnetic properties of Ni. (After Bozorth, R. M., *Ferromagnetism*, Van Nostrand, New York, 1956, 440. With permission.)

An explanation of this behavior may be obtained by examining the band structure of the alloys. The d band can contain up to 10 electrons in an energy range of about 3 eV. This represents a high density of states. The s band can accept two electrons in a range of more than 7 eV; this is a much lower density than that of the d band. Since these elements can be described by a rigid-band model, the probability of finding an added electron will vary directly with the density of states. There is a high probability that the s or p valence electrons of alloying elements, in substitutional solid solution in the lattice of the transition element, will be found in the d levels of the transition element. Ni-Cu alloys furnish a good example of this behavior. The replacement of a Ni ion by a Cu ion adds one electron to the system. This electron will most probably occupy a d state. This diminishes the number of unbalanced spins of the Ni ions and decreases the magnetization. As increasing numbers of Cu ions go into solution the number of unbalanced d-level spins decreases. A point is reached at which the electrons from the copper ions fill the d levels of the Ni ions to a maximum extent. This occurs near 60 at.% Cu-40 at.% Ni (see Sections 7.8.4 and 7.13.2). At this composition the spin imbalance is effectively zero and the alloy shows zero effective Bohr magnetons per atom, as shown in the figure. This tendency would be expected to increase as the number of valence electrons of the solute atom increases. This is indeed the case. The number of valence electrons contributed by each solute element is shown in parentheses in Figure 9-17. Compositions representing nonmagnetic alloys decrease with increasing valence of the added element.

On the other hand, Pd, which has an electronic configuration very similar to that of Ni, shows almost no effect over a wide range of compositions. These elements form a continuous series of solid solutions, but the change in the lattice parameter could account for the small decrease in the spontaneous magnetization (see Figure 9-11). It also will be noted that Mn, which adds holes, has a positive magnetic effect since it adds to the spin imbalance of the d levels.

Iron-base, substitutional, solid solution alloys with nontransition elements show different behavior than that of the Ni-base alloys. These ions in solution in the iron lattice result in a decrease of about 2.2 μ_B for each such alloying ion. On this basis, no electron interactions occur between the Fe and alloy ions. The valence electrons from the solute ions do not interact with the iron d levels as in the case of the Ni-base alloys; therefore, the rigid-band model does not apply. These electrons could occupy a separate valence band. In contrast to this, Fe-base alloys with other transition elements having 8 or more d electrons have properties in reasonable agreement with the rigid-band model.

The Co-Ni system forms a continuous series of solid solutions, all of which are magnetic. Alloys up to about 30 at.% Ni show ordered structures at room temperature. This has no effect upon T_c as a function of composition, since no magnetic discontinuities appear. The number of magnetons of a "Co-Ni ion" is given by $n_B = 0.57\ C_{Ni} + 1.71\ (1 - C_{Ni})$ where C_{Ni} is the atom fraction of Ni ions in the alloy. In other words, the spin imbalances are additive.

The Fe-Co system is more complex at room temperature than the alloys considered previously. The Curie temperature is a complicated function of composition and an ordered lattice structure occurs at the FeCo composition. The composition close to $Fe_{0.7}Co_{0.3}$ has $2.5\mu_B$ per ion, the largest thus far observed.

The Fe-Ni system shows a greater variation in T_c as a function of composition than any of the foregoing systems. Both Ni and Co increase the magnetization when in solution in Fe; the mechanism involved in this is not known as yet.

These effects are shown for several alloy systems in Figure 9-18. Here, the magnetization is plotted against the total number of electrons per atom of the alloy instead of atom percent. This device permits the presentation of a larger number of systems without considering configurational details.

Again, as discussed above, the added elements either diminish or increase the spin imbalance of the d levels and cause either a decrease or an increase in the magnetization of the resultant alloys.

Many other alloys of the transition elements show magnetic behaviors. These include alloys (intermetallic compounds) such as Fe_3Al, Au_4V, Au_4Mn, FeAl, and Cu_5Mn in which the magnetic behavior is a strong function of the degree of perfection of the two sublattices composing the ordered structure of the compounds. Some of these

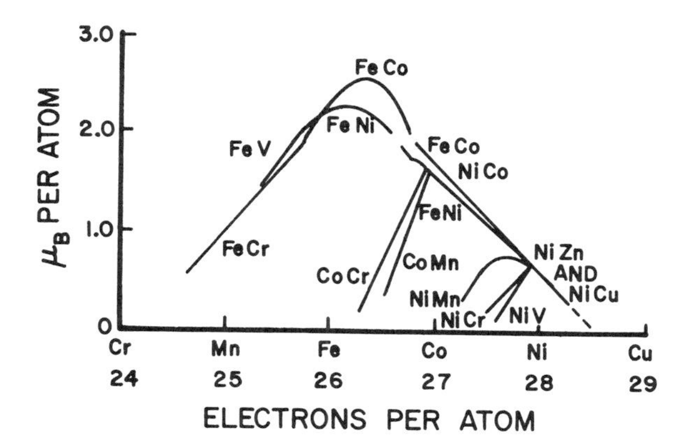

FIGURE 9-18. Number of Bohr magnetons per atom as a function of the total number of electrons per atom in an alloy. (After Bozorth, R. M., *Ferromagnetism,* Van Nostrand, New York, 1951, 440. With permission.)

alloys, one component of which has no magnetic moment, show that clusters of ions with "giant moments" are present. This is the case for FeAl where the concentration of the clusters decreases with increasing degrees of ordering.

Some additional systems which contain ferromagnetic alloys include Mn-Al, Mn-As, Mn-Bi, Mn-N, Mn-P, Mn-Sb, Cr-Pt, Cr-S, Cr-Te, Co-Pt, and Fe-Pt. While most of these systems do not contain a ferromagnetic component, they do include the elements Mn and Cr. In these cases it appears that the magnetic alloys are such that the ratios r_a/r_d result in positive values of J_e which are responsible for ferromagnetic properties. This also holds for the Au_4V alloy noted earlier (see Section 9.5 and Figure 9-11). It will be recalled that λ is a function of J_e (Equation 9-41) and that the molecular field, in turn, is a function of λ (Equation 9-1). Thus, the Weiss field is a function of r_a/r_d. This accounts for the magnetic alloys in those systems noted above which contain no ferromagnetic components.

9.11. HEUSLER ALLOYS

Heusler (1898) first demonstrated that alloys of Cu-Mn-Sn and Cu-Mn-Al showed spontaneous ferromagnetic behavior similar to that of nickel. These alloys are close to the compositions given by Cu_2MnSn and Cu_2MnAl. Other alloys, including such compositions as Cu_2MnIn and Cu_2MnGa, have similar properties. Similar compositions in silver-base alloys also show this behavior.

Alloys of this type are intermetallic compounds and all have a similar ordered lattice type. The Mn and Sn, or Mn and Al, atoms take body-centered sites within a face-centered lattice consisting of the Cu atoms, forming sublattices (this structure is shown in Figure 9-19).

The properties of this class of alloys are highly dependent upon the degree of perfection of the ordering, or superlattice. Anything that disturbs this array diminishes the net magnetic moment and lowers the Curie temperature. Such factors as compositional variations and stress can cause this. For example, quenched alloys are stressed and largely disordered and show less desirable magnetic properties than in the unstressed condition. When these quenched alloys are aged (ordered and stress relieved) in the neighborhood of 200°C for increasing times, their properties improve with the time of the treatment. Their degree of crystallographic perfection increases with the aging treatment.

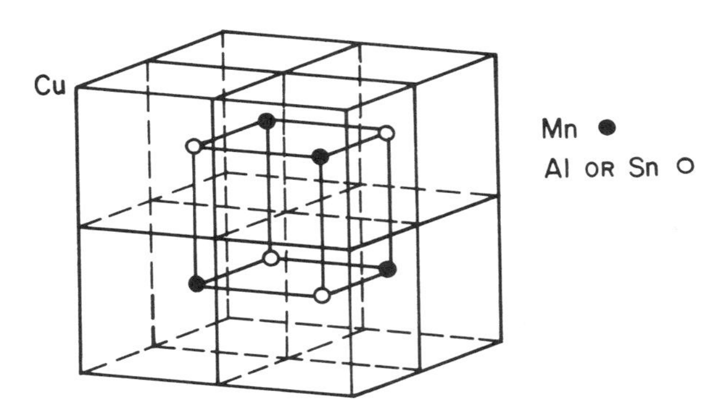

FIGURE 9-19. Superlattice of Heusler-type alloys. (After Bozorth, R. M., *Ferromagnetism*, Van Nostrand, New York, 1951, 329. With permission.)

In addition to crystallographic order, the distance between the Mn ions appears to be critical. The role of the alloying seems to be one of maintaining the Mn ions sufficiently far apart so that their d levels do not overlap. An empirical limit (at variance with Slater's calculation, Section 9.5) for phases with direct exchange coupling appears to be a minimum distance of 2.8 Å for Mn atoms. MnAs and MnSb, with lattice spacings (c/2) of 2.84 and 2.89 Å, respectively, are ferromagnetic (a similar limit for Cr ions is 3.05 Å). Alloy combinations of this class must be such that the interionic distances for Mn must be above this limit in order that the exchange integral can become positive and the exchange energy negative.

This model predicts that both the exchange energy and the Curie temperature should reach a maximum at a given ratio of r_a/r_d. Empirical plots of the Curie temperatures of various binary and ternary alloys as a function of r_a/r_d show that this maximum occurs at a value of r_a/r_d, between 1.8 and 1.9. The exchange coupling increases sharply, as measured by T_c, from $r_a/r_d \simeq 1.5$ up to the maximum which appears between 1.8 and 1.9. Thereafter, the exchange rapidly diminishes up to a value of about 2.1. It thus appears that the influence of alloying has an effect upon these properties similar to that shown in Figure 9-11 for the pure elements.

Many other intermetallic compounds with Mn have since been shown to have ferromagnetic properties. These include MnAl, MnAs, MnB, MnBi, Mn_4N, MnP, Mn_2Sb, MnSb, and Mn_4Sn. Other compounds based upon chromium show weaker ferromagnetic behavior.

9.12. FERRIMAGNETISM

Magnetite, Fe_3O_4, has long been known for its magnetic properties. Its formula also may be expressed as $Fe^{+2}Fe_2^{+3}O_4$. It has been shown that materials with good magnetic properties may be obtained when the divalent ion is replaced by Mn, Co, Ni, Cu, Mg, Zn, or Cd. The trivalent ion may be replaced by Mn, Co, Al, or Ga. The nonmetallic divalent ion can be S or Se, but is usually oxygen. These have complex cubic lattices and are called "spinels". Spinels of the form $M^{+2}Fe_2O_4$ are called "ferrites". "Mixed" ferrites can be made by using several divalent metals for M^{+2}. These materials form crystal structures in which the magnetic moments of the sublattices are opposite to each other. A net magnetic moment results when those of the sublattice are unequal (Figure 9-20a). Other ferrites have complex hexagonal lattices, as do the barium ferrites. The other principal types of compounds showing ferrimagnetic properties are

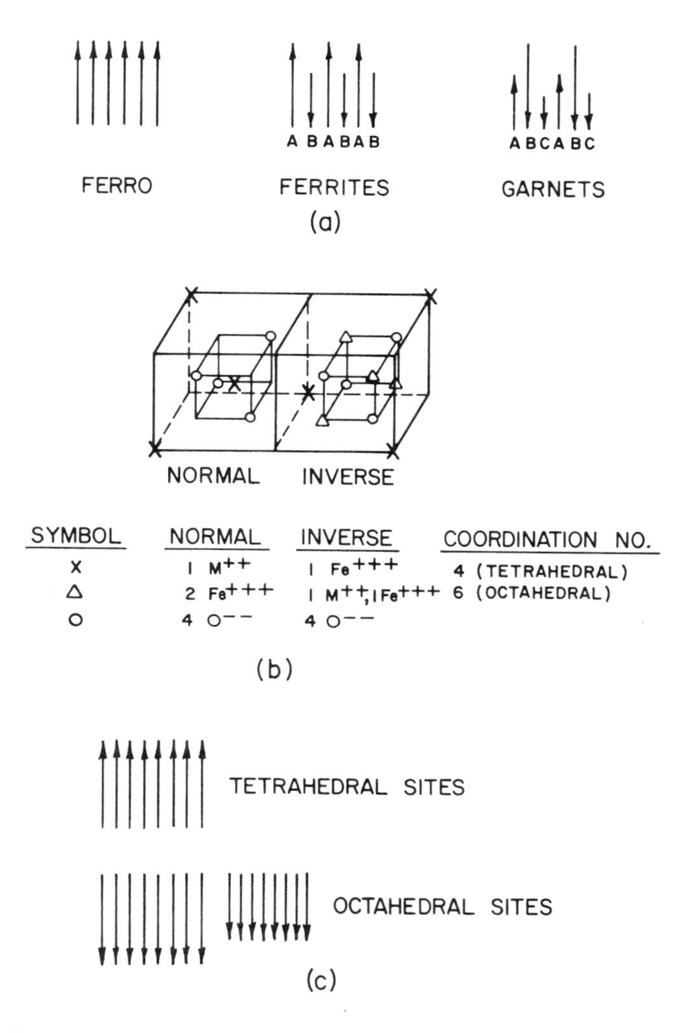

FIGURE 9-20. (a) Models of spin configurations; (b) structures of normal and inverse ferrites (After Bozorth, R. M., *Ferromagnetism*, Van Nostrand, New York, 1951, 224. (c) Spin array for magnetite. (After Kittel, C., *Introduction to Solid State Physics*, 3rd ed., John Wiley & Sons, New York, 1966, 472. With permission.)

garnets. These are of the composition given by $M_3Fe_5O_{12}$, where M is one of the trivalent rare earths. Almost all of these materials are poor electrical conductors; the garnets are insulators.

As is the case for the Heusler-type alloys, the crystalline order is very important in ferrites. The unit cell contains 32 oxygen atoms in a close-packed cubic array with the divalent and trivalent ions in the interstices. The divalent ions occupy the tetrahedral sites and the trivalent ions, the octahedral sites. The unit cell contains 16 trivalent ions and 8 divalent ions. In the "normal" structure, the divalent ions are in the tetrahedral sites and the trivalent ions are in the octahedral sites (Figure 9-20b). In the "inverse" structure the divalent ions take up octahedral sites and the trivalent ions are equally in tetrahedral and octahedral positions. These crystal arrays constitute assemblies of two sublattices. The resultant magnetic moment is the algebraic sum of the moments of the sublattices. The magnetic moment of ferrite, thus, is the net imbalance of the magnetic moments of the sublattices. This is shown in Figure 9-20c. Here it can be seen that the moments of the trivalent Fe ions in both positions nullify each other. The net magnetic moment arises from the divalent Fe ions.

In zinc ferrite, $ZnFe_2O_4$, the Zn ions are located on the tetrahedral sites. Zinc, with a filled 3d band, has a zero magnetic moment, so the magnetization of the A sublattices M_A, will be zero. The trivalent Fe^{+3} ions occupy the octahedral sites. These ions have exactly half-filled 3d bands. Thus, the magnetization of the B sublattice, M_B, will be 5 μ_B/ion. This gives a saturation magnetization of $5\mu_B$ for each molecule.

Nickel ferrite, $NiFe_2O_4$, has the inverse structure. The Fe^{+3} ions occupy the octahedral sites and cancel the magnetic effects of those on the tetrahedral sites. The remaining octahedral sites are occupied by Ni^{+2} ions, each with 2 μ_B/atom. So, $M_B = 2\mu_B$. This gives saturation magnetization of 2 μ_B/molecule (See Figure 9-20).

The common normal spinels are $CdFe_2O_4$ and $ZnFe_2O_4$. The common inverse spinels are formed by the divalent ions Mn, Fe, Co, Ni, Cu, and Mg. Since these are cubic systems, they show cubic anisotropy, the easy directions being [110] or [111].

The spontaneous magnetization of many ferrimagnetic material approaches that of nickel at low temperatures. These are of the order of 3-4μ_B/formula unit. Some of the garnets approach 30μ_B/formula unit. Garnets usually contain a rare-earth ion in addition to those present in ferrites, and three sublattices are formed.

The magnetic ordering of these materials involves at least two sublattices which have their magnetic moments in opposite directions. These moments arise from the spontaneous magnetizations of the sublattices.

The curves of magnetization vs. temperature of many ferrites are similar in shape to those of ferromagnetic materials. Others, such as some of the rare-earth containing garnets, drop steeply, between 0 and 300 K, to nearly zero and then go through a maximum before they reach a Curie temperature of about 570 K. Magnetization curves can take on other shapes as shown in Figure 9-21. This is given in terms of the magnetizations of the two sublattices, M_A and M_B. The resultant, given by the dotted line in the figure, is the magnetization of the ferrite. Many other shapes are possible, depending upon the curve shapes of the magnetizations of the sublattices.

The foregoing suggests that the properties of ferrites result from a ''negative'' rather than a cooperative process. This was explained by Néel (1948) on the basis of the sublattices noted previously. His approach is outlined here. Consider the solid to be made up of two cubic sublattices A and B. Not only will there be A-B interactions such as shown in Figure 9-20, but A-A and B-B influences as well. It is assumed here that interactions other than between nearest neighbors are negligible.

The resultant interaction on a given ion depends upon the degree of alignment of its neighbors. The ion thus behaves as though it is in a molecular field λM, where λ is the field constant. This can be extended by considering that an A ion reacts with the B sublattice (A-B) and also with the Weiss field of its own sublattice (A-A). An ion in the B sublattice is also interacting with the A sublattice (B-A) and its own Weiss field (B-B). If M_A and M_B are the magnetizations of the two sublattices, and $-\alpha$, $-\beta$, $-\gamma$ and $-\delta$ represent the A-A, A-B, B-A, and B-B respective field constants, then the molecular field affecting an A ion is, for spontaneous magnetization,

$$H_A = -\alpha M_A - \beta M_B \qquad (9\text{-}57a)$$

and that affecting a B ion is

$$H_B = -\gamma M_A - \delta M_B \qquad (9\text{-}57b)$$

In the presence of an external field, the molecular fields of the two sublattices become

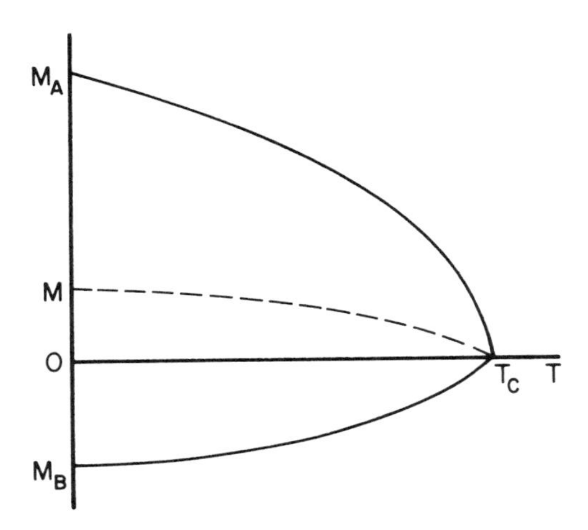

FIGURE 9-21. Resultant magnetization for two sublattices, A and B, for x = 1, Equation 9-60.

$$H_A = H - \alpha M_A - \beta M_B \tag{9-58a}$$

and

$$H_B = H - \gamma M_A - \delta M_B \tag{9-58b}$$

The equations for the molecular fields of the two sublattices given above may be used in Equation 8-112 to give their respective magnetizations. This gives, using the notation B_S instead of B_J

$$M_A = Ng\mu_B SB_S \left[\frac{Sg\mu_B H_A}{k_B T}\right] \tag{9-59a}$$

and

$$M_B = Ng\mu_B SB_s \left[\frac{Sg\mu_B H_B}{k_B T}\right] \tag{9-59b}$$

when Equations 9-58a and 9-58b are used. The total magnetization is

$$M = M_A + M_B = Ng\mu_B S \left\{ B_S \left[\frac{Sg\mu_B H_A}{k_B T}\right] + B_S \left[\frac{Sg\mu_B H_B}{k_B T}\right] \right\} \tag{9-60}$$

Solutions to this equation give curves such as are shown in Figure 9-21. As would be expected, the total magnetization is strongly influenced by the ions composing the sublattices. As a result of this, it is possible to obtain curves of total magnetization vs. temperature whose shapes are much more complex than that given in Figure 9-21.

The spontaneous magnetizations of these materials vanish at the Curie temperature, T_c. They are paramagnetic at temperatures above T_c. Here, it is assumed that each sublattice obeys the Curie-Weiss law (Equation 9-8).

9.13. ANTIFERROMAGNETISM

Néel and others have shown that a substance could have considerable magnetic or-

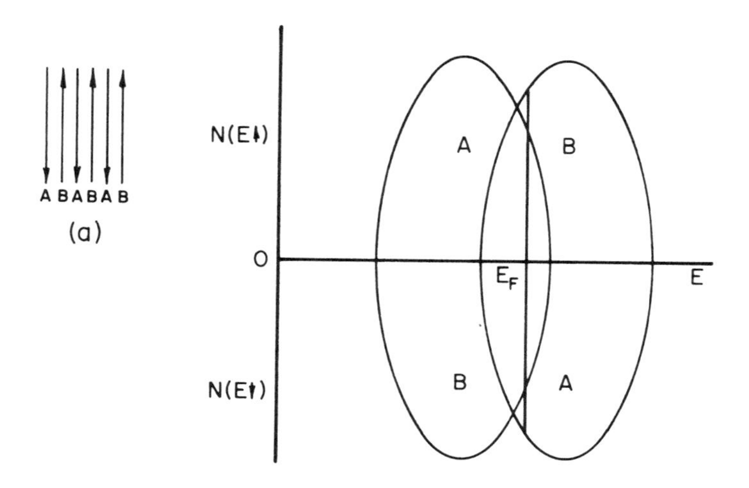

FIGURE 9-22. (a) Spin array on two sublattices of an antiferromagnetic material; (b) d-subbands of an antiferromagnetic transition element such as Mn or Cr.

dering within its component sublattices and still show little magnetization or susceptibility because the spins are opposed and J_e is negative (Figure 9-22). These materials can show an increasing but small degree of magnetization frequently at relatively low temperatures. As the temperature is increased a point is reached at which they become paramagnetic. The temperature at which paramagnetic behavior begins is the Néel temperature, T_N. Thus, a plot of susceptibility vs. temperature shows a peak at T_N.

As in the model for ferrimagnetism, the crystal structure of an antiferromagnetic material has been shown to be compose of at least two sublattices. For simplicity, assume two sublattices to be present. The parallel spins on each of the sublattices oppose and cancel each other at 0 K and the magnetization is zero. This interaction of spins is very strong at low temperatures, but becomes slightly less balanced and small magnetizations are induced by applied fields. The interaction between the sublattices diminishes as the temperature is increased and the magnetic susceptibility increases. A temperature is finally reached at which the thermal forces are greater than the exchange forces and the spins become random. This is the Néel temperature, T_N. The substance shows paramagnetic behavior at temperatures above T_N.

First consider the behavior of these materials at temperatures well above T_N. Here their paramagnetic behavior is given by Equation 8-50. This can be used to describe the magnetizations of the two sublattices A and B as

$$M_A = \frac{N\mu_M^2}{3k_BT} H_A \tag{9-61a}$$

and, assuming an equal number of similar moments on the B sublattice,

$$M_B = \frac{N\mu_M^2}{3k_BT} H_B \tag{9-61b}$$

where, from Equation 8-103,

$$\mu_M^2 = g^2S(S+1)^2\mu_B^2 \tag{9-61c}$$

The Weiss fields on an ion in each sublattice are (compare with Equation 9-58):

$$H_A = H - \alpha M_A - \beta M_B \tag{9-62a}$$

and

$$H_B = H - \beta M_A - \alpha M_B \tag{9-62b}$$

in which H is the applied field and α and β are the Weiss constants. The constant α represents the A-A and B-B interactions while β is the constant for the unlike, antiferromagnetic, A-B sublattice interactions.

Summing Equations 9-61a and 9-61b

$$M = M_A + M_B = \frac{N\mu_M^2}{3k_BT}[H_A + H_B] \tag{9-63}$$

Substituting Equations 9-62a and 9-62b into 9-63

$$M = \frac{N\mu_M^2}{3k_BT}[2H - (\alpha + \beta)M_A - (\alpha + \beta)M_B]$$

or

$$M = \frac{N\mu_M^2}{3k_BT}[2H - (\alpha + \beta)(M_A + M_B)] \tag{9-64}$$

$$= \frac{N\mu_M^2}{3k_BT}[2H - (\alpha + \beta)M]$$

since $M = M_A + M_B$. Assuming M and H are parallel, Equation 9-64 can be reexpressed as

$$M = \frac{2N\mu_M^2}{3k_BT}H - \frac{(\alpha + \beta)N\mu_M^2}{3k_BT}M$$

and rearranged as

$$M\left[1 + \frac{(\alpha + \beta)N\mu_M^2}{3k_BT}\right] = \frac{2N\mu_M^2}{3k_BT}H$$

This expression is solved for M to obtain

$$M = \frac{\dfrac{2N\mu_M^2H}{3k_BT}}{\dfrac{3k_BT + (\alpha + \beta)N\mu_M^2}{3k_BT}} = \frac{2N\mu_M^2H}{3k_BT + (\alpha + \beta)N\mu_M^2} \tag{9-65}$$

The susceptibility is obtained from Equation 9-65 as

$$\chi = \frac{M}{H} = \frac{2N\mu_M^2}{3k_BT + (\alpha + \beta)N\mu_M^2} \tag{9-66}$$

Then by dividing the numerator and denominator by $3k_B$, the susceptibility is found to be

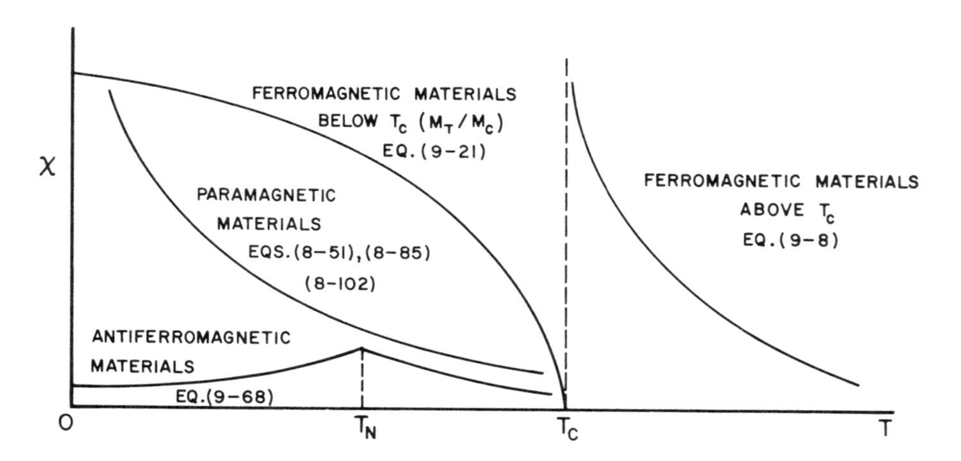

FIGURE 9-23. Schematic comparison of magnetic properties.

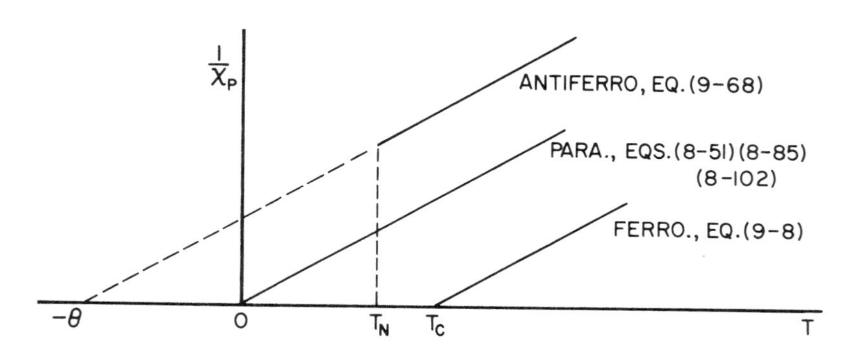

FIGURE 9-24. Reciprocal paramagnetic susceptibilities of magnetic materials (From Dekker, A. J., *Solid State Physics*, Prentice-Hall, Englewood Cliffs, N.J., 1957, 485. With permission.)

$$\chi = \frac{\dfrac{2N\mu_M^2}{3k_B}}{T + \dfrac{(\alpha + \beta)N\mu_M^2}{3k_B}} \qquad (9\text{-}67)$$

Equation 9-67 can be rewritten in more general form as

$$\chi = \frac{C}{T + \theta} \qquad (9\text{-}68)$$

in which the Curie constant is

$$C = \frac{2N\mu_M^2}{3k_B} = \frac{2Ng^2S(S+1)^2\mu_B^2}{3k_B} \qquad (9\text{-}68a)$$

and the "paramagnetic temperature" is given by

$$\theta = \frac{(\alpha + \beta)N\mu_M^2}{3k_B} \qquad (9\text{-}68b)$$

The molar susceptibilities of most of these materials is of the order of about 10^{-3},

Table 9-4
ANTIFERROMAGNETIC CRYSTALS

Substance	Paramagnetic ion lattice	Transition temperature T_N (K)	Curie-Weiss Θ (K)	$\frac{\Theta}{T_N}$	$\frac{\chi(0)}{\chi(T_N)}$
MnO	FCC	116	610	5.3	2/3
MnS	FCC	160	528	3.3	0.82
MnTe	Hexagonal layer	307	690	2.25	—
MnF_2	BCT	67	82	1.24	0.76
FeF_2	BCT	79	117	1.48	0.72
$FeCl_2$	Hexagonal layer	24	48	2.0	<0.2
FeO	FCC	198	570	2.9	0.8
$CoCl_2$	Hexagonal layer	25	38.1	1.53	—
CoO	FCC	291	330	1.14	—
$NiCl_2$	Hexagonal layer	50	68.2	1.37	—
NiO	FCC	525	∿2,000	∿4	—
Cr	BCC	308	—	—	—

From Kittel, C., *Introduction to Solid State Physics*, 5th ed., John Wiley & Sons, New York, 1976, 483. With permission.

below T_N. The similarity between Equation 9-68 and that for the behavior of ferromagnetic materials above T_c (Equation 9-8 the Curie-Weiss law) will be recognized. In the present case the parameter θ in the denominator has a positive sign and is not the Néel temperature. In the Curie-Weiss law, the corresponding parameter is the Curie temperature, T_c, which is positive and has a negative sign. Schematic diagrams of the behaviors of paramagnetic, ferromagnetic, and antiferromagnetic materials are given in Figures 9-23 and 9-24.

Figure 9-24 shows the reciprocal susceptibility for these types of materials as functions of temperature (some experimental data are given in Table 9-4).

The behavior of antiferromagnetic materials at temperature below T_N can be obtained starting with Equations 9-61 and 9-62, and recalling that H = 0,

$$M_A = -\frac{N\mu_M^2}{3k_BT}(\alpha M_A + \beta M_B)$$

or

$$M_A\left[1 + \frac{N\mu_M^2\alpha}{3k_BT}\right] + \frac{N\mu_M^2}{3k_BT}\beta M_B = 0 \tag{9-69}$$

and in the same way

$$M_B\left[1 + \frac{N\mu_M^2\alpha}{3k_BT}\right] + \frac{N\mu_M^2}{3k_BT}\beta M_A = 0 \tag{9-70}$$

Equations 9-69 and 9-70 can be equated and solved for $T = T_N$, using Equation 9-68a, to find

$$T_N = \frac{C(\beta - \alpha)}{2} \tag{9-71}$$

From Equation 9-71 it can be seen that T_N will rise as the interactions between the two sublattices (A-B) increases because of the increasing degree of antiferromagnetic behavior which must result. By the same token, T_N will decrease if the interactions within the like lattices (A-A and B-B) increase.

A relationship can be obtained for T_N and θ starting with Equations 9-68a and 9-68b.

$$\theta = \frac{(\alpha + \beta) N \mu_M^2}{3k_B} = \frac{(\alpha + \beta) C}{2} \quad (9\text{-}72)$$

Equation 9-72 is divided by Equation 9-71 to give

$$\frac{\theta}{T_N} = \frac{(\alpha + \beta) C}{2} \cdot \frac{2}{C(\beta - \alpha)} = \frac{\alpha + \beta}{\beta - \alpha} \quad (9\text{-}73)$$

Experimental data for some antiferromagnetic compounds are given in Table 9-3. The fact that the Néel temperature is less than the paramagnetic temperature, θ, is significant. It means that each sublattice must possess an individual Weiss constant and that fields of each must oppose each other. This must be the case since α must be positive for $T_N < \theta$ to occur. However, a limit exists for the ratio given by Equation 9-73. This limit depends upon the magnitude of α with respect to β. When α becomes larger, the next-nearest neighbors in a sublattice tend to become antiparallel instead of parallel to the given reference ion. The description given by Equation 9-73 is not valid when this occurs. In this event, four sublattices are required to explain the properties of the substance.

At temperatures below the Néel temperature two limiting situations may occur in single crystals. One takes place where the magnetic field is perpendicular to the spin orientation. The other occurs where the field is parallel to the spin orientation. The following discussions of these two conditions include the simplifying assumptions that the crystalline anisotropy causes the spins to lie in easy directions and that only the mutual reaction between the two sublattices (A-B) need to be taken into account.

The negative value of J_e which is responsible for the tendency of the spins to oppose each other is partially overcome by the torque induced by the field (Figure 9-25a and b). These are now at some angle ϕ with respect to their original orientation. The molecular fields of the sublattices oppose this rotation. The forces acting on the spins of sublattice A are H and $-\beta M_A$. Those acting on the spins of the other sublattice are assumed to be equal at equilibrium. Thus, for small angles,

$$2\beta M_A \phi = H \quad (9\text{-}74)$$

For the present case, where $M_A = M_B$,

$$M = (M_A + M_B)\phi = 2 M_A \phi \quad (9\text{-}75)$$

Dividing Equation 9-75 by 9-74 the susceptibility perpendicular to the spin is

$$\chi_\perp = \frac{M}{H} = \frac{2M_A\phi}{2\beta M_A \phi} = \frac{1}{\beta} \quad (9\text{-}76)$$

This is a constant and independent of temperature. Additionally, it can be shown that Equation 9-76 also holds when A-A and B-B interactions are included in the model.

The susceptibility where the field is parallel to the spin orientation is considered next. At 0 K the spins of the two sublattices are equal and opposite. The susceptibility is, thus, equal to zero. As previously noted, as the temperature increases this spin

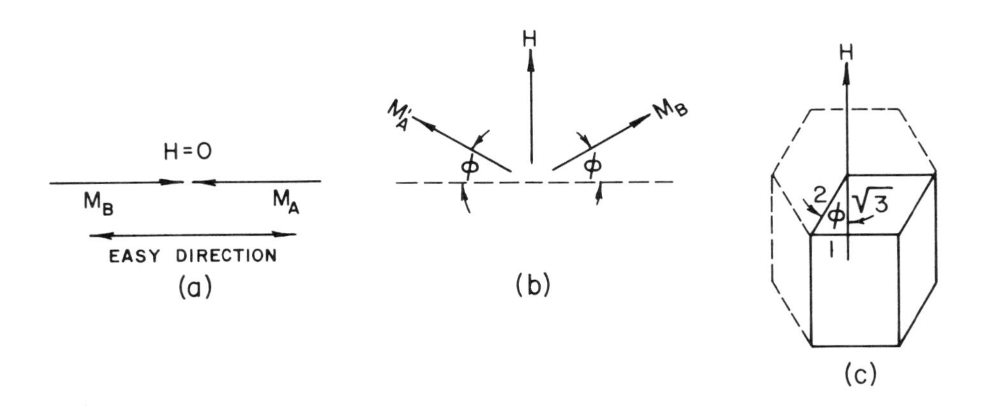

FIGURE 9-25. (a) Antiferromagnetic material in the absence of a field; (b) field applied perpendicular to the easy direction; (c) field applied perpendicular to a prism plane of a hexagonal lattice.

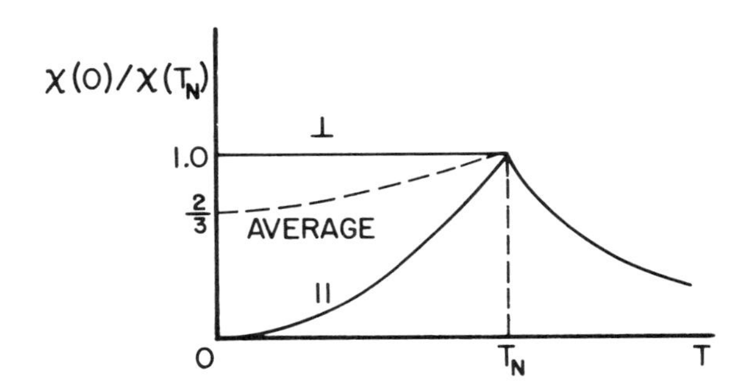

FIGURE 9-26. Magnetic susceptibility as functions of orientation and temperature.

balance diminishes in a uniform way and the susceptibility increases smoothly, up to T_N. At T_N, $\chi_\perp = \chi_\parallel$. This is shown schematically in Figure 9-26. Paramagnetic behavior is observed at and above T_N.

An approximation may be made for the susceptibility of materials with complex lattice types, if the components relative to the easy direction are considered. The susceptibility may be written as

$$\chi = \frac{M}{H} = \frac{MH}{H^2} \simeq \frac{M(H_\parallel + H_\perp)}{H^2}$$

The fields parallel and perpendicular to the easy direction are

$$H_\parallel = H \cos \phi \quad \text{and} \quad H_\perp = H \sin \phi$$

So that H vanishes from the numerator upon substitution and

$$\chi \simeq \frac{M(H \cos \phi + H \sin \phi)}{H^2} = \frac{M(\cos \phi + \sin \phi)}{H}$$

The corresponding components of M are introduced to give

$$\chi \simeq \frac{M_\parallel \cos \phi + M_\perp \cos \phi}{H}$$

The susceptibilities for the two directions are

$$\chi_{\parallel} = \frac{M_{\parallel}}{H_{\parallel}} = \frac{M_{\parallel}}{H \cos \phi} \quad \text{and} \quad \chi_{\perp} = \frac{M_{\perp}}{H_{\perp}} = \frac{M_{\perp}}{H \sin \phi}$$

These are rearranged to give equations for $M_{\parallel}$ and $M_{\perp}$:

$$M_{\parallel} = \chi_{\parallel} H \cos \phi \quad \text{and} \quad M_{\perp} = \chi_{\perp} H \sin \phi$$

The substitutions of the magnetizations of the two directions into the equation for χ results in

$$\chi \simeq \chi_{\parallel} \cos^2 \phi + \chi_{\perp} \sin^2 \phi$$

since H vanishes from the equation.

This relationship is useful in dealing with polycrystalline materials. However, in order to do this, an average value of ϕ must be used because of the random orientation of the grains. This is obtained from an approximation based upon the Langevin function (Equation 8-47). It gives, for large angles, $\langle\phi\rangle \simeq \theta - 1/\theta$ where θ is the maximum angle of the range. Here $\theta = \pi/2$, so that $\langle\phi\rangle \simeq \pi/2 - 2/\pi \simeq 54°$. The use of simple fractions to approximate $\cos^2 \langle\phi\rangle$ and $\sin^2 \langle\phi\rangle$ results in

$$\chi \simeq \chi_{\parallel}/3 + 2\chi_{\perp}/3$$

An evaluation of this equation may be made by comparing the susceptibility at 0 K, $\chi(0)$, with that at the Néel temperature, $\chi(T_N)$. At 0 K, $\chi(0) = 2/3\chi_{\perp}$, since $\chi_{\perp}$ is a constant and independent of temperature and, $\chi_{\parallel} = 0$ as have been shown previously. At T_N, $\chi_{\parallel} = \chi_{\perp}$ so that $\chi(T_N) = \chi_{\perp}$. The ratio of the susceptibilities obtained in this way is

$$\frac{\chi(0)}{\chi(T_N)} \simeq \frac{2/3\chi_{\perp}}{\chi_{\perp}} = \frac{2}{3}$$

This result is in reasonable agreement with the data in Table 9-4 for the FCC and BCT materials, despite the simplifying assumptions. The general behavior is shown in Figure 9-26.

Another approximation may be made for the case of materials with hexagonal layered structures. Here it is assumed that the susceptibility parallel to the c axis is negligible for a layered structure and that the easy directions are parallel to the a axes in the basal plane. When the field is perpendicular to a prism plane of such a structure cell, then $\sin \phi = 1/2$ and $\cos \phi = \sqrt{3}/2$, Figure 9-25c. These conditions result in

$$\chi \simeq 3/4\chi_{\parallel} + 1/4\chi_{\perp}$$

By means of the same method of evaluation used previously, the ratio of the susceptibilities is

$$\frac{\chi(0)}{\chi(T_N)} \simeq \frac{1/4\chi_{\perp}}{\chi_{\perp}} = 1/4$$

The simplifying assumptions leading to the susceptibility ratio $\simeq 1/4$ could explain the low value given for $FeCl_2$ in Table 9-4, if the specimens were single crystals or if they were polycrystalline materials in which the grains had a strong preferred orientation. The same holds for CrSb, another compound with a hexagonal layered structure.

9.14. MAGNETIC MATERIALS (COMMERCIAL)

A brief review of the engineering parameters and their units (previously discussed in Section 9.8) is given here for convenience. A typical magnetization curve is given in Figure 9-15. Here B is the magnetic flux density, or induction, in gausses, and H is the magnetizing field in oersteds. B_r is the remanence or residual magnetization (also called residual induction) that remains when the magnetizing field, H, is removed. B_s is the saturation magnetization. The value of H_c provides a measure of the demagnetizing force required to cause the residual magnetization to go to zero. The area within the curve is a measure of the expended energy.

Magnetically soft materials have high permeabilities and a high value of B_s, since $\mu = B/H$. It is also desirable that these materials will absorb a minimum of energy. This is the same as saying that the area enclosed by the hysteresis loop will be as small as possible; H_c will be as small as possible and approach zero. Magnetically hard materials (permanent magnets) must show diametrically opposite behaviors. In this case it is desirable to have a large hysteresis loop with large values for both B_r and H_c, in order that the magnetization be large and "permanent".

The portion of the B-H curve in the second quadrant of Figure 9-15, between B_r and H_c, is called the demagnetization curve. It is important in the selection of permanent magnetic materials. These may be exposed to other magnetic fields during their service lives and have their magnetizations inadvertently reduced.

The energy product is a more specific criterion for the selection of magnetic materials. This is found for a given point on the B-H curve by the product of B and H. The energy product curve for a given material may be obtained when this is done for a sufficient number of points taken from the B-H curve. The maximum energy product, $(B_dH_d)_{max}$, is considered by many engineers to be the best individual basis for the comparison and selection of permanent magnet materials. However, it is most prudent to consider both the demagnetization curve, giving B_r and H_c, and the maximum energy product (both curves are shown in Figure 9-27).

9.14.1. Permanent Magnet Alloys

The early permanent magnets were made of high-carbon steels in the quenched-and-tempered condition. Later, steels containing tungsten, chromium, and cobalt were used (see Table 9-5). These are given to show the effects of alloying constituents and to serve as a basis for comparison with other types of alloys. The effects of alloying are apparent from the table. The plain-carbon steels, now obsolete for magnetic purposes, have been included for purposes of comparison with the other ferrous alloys. It will be noted that the values of B_r are nearly constant. However, the coercive forces and the maximum energy products increase considerably, indicating magnetic materials of increasing strength and magnetic permanence as the alloy content increases.

Magnet steels, such as those given in the table, are not used at temperatures above 100°C because of the metallurgical changes which occur in steels upon tempering. The resulting changes in metallographic structure are accompanied by decreases in magnetic properties. When such steels are maintained in the as-quenched condition, an aging effect occurs at room temperature. Some decrease of the energy product occurs. The largest portion of this change takes place directly after quenching and becomes very small in a few weeks time. The magnetic properties of these steels also can be diminished by mechanical shock. However, after repeated shocks, the properties approach some lower, approximately stable, values. Thermal and mechanical treatments have been used to minimize these tendencies toward instability.

The carbon-free alloys (Table 9-6) represent only a few of the commercial alloys. They are much less susceptible to aging (when properly treated) than are the steels.

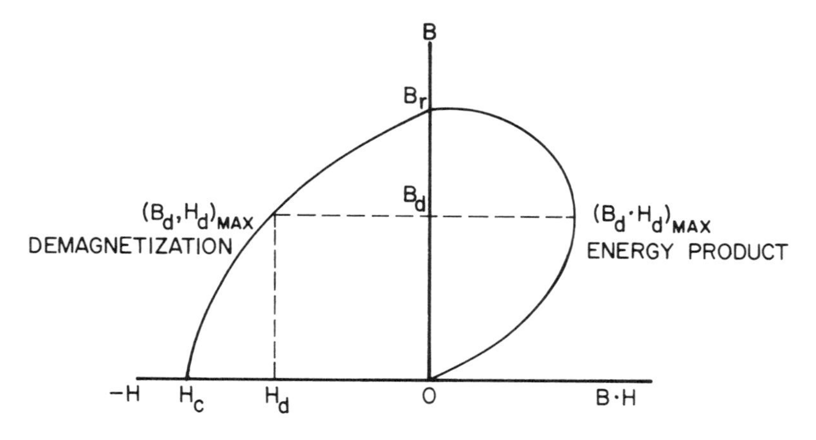

FIGURE 9-27. Schematic curves for the demagnetization and energy product of a permanent magnet.

Table 9-5
TYPICAL PROPERTIES OF SOME MAGNET STEELS

Materials	Nominal composition (Wt %) C	Mn	Cr	W	Co	Remanence B_r (G)	Coercive force H_c (Oe)	Max energy product $(B_dH_d)_{max}$ (millions)
Carbon steel	0.65	0.85	—	—	—	10,000	42	0.18
	1.00	0.50	—	—	—	9,000	51	0.20
Chromium steel	1.00	0.50	3.50	—	—	9,500	66	0.29
Tungsten steel	0.70	0.50	0.50	6.00	—	9,500	74	0.33
Cobalt steel	0.70	0.35	2.50	8.25	17.00	9,500	170	0.65
	0.80	0.55	5.75	3.75	36.00	9,750	240	0.93

From Lyman, T., Ed., *Magnetic, Electrical and Other Special-Purpose Materials* Vol. 1, 8th ed., American Society for Metals, Metals Park, Ohio, 1961, 789. With permission.

Remalloy, which was the first of this class of alloys to be produced, is representative of the Fe-Co-Mo alloys. These are dispersion hardened and can be forged and rolled. The magnetic properties of this class of alloys are intermediate between those of the steels and of the Alnico alloys.

Two other members of this intermediate group, Cunife and Cunico, are precipitation hardening and can be worked mechanically. This permits their fabrication into sheets, thin tapes, and wires. The best properties of Cunife are obtained after successive cold reductions totalling about 95% reduction in area when intermediate and final thermal treatments are used. Such treatments can approximately double the energy product of Cunife. The increases in B_r, H_c, and the energy product are anisotropic in sheet and tape, where the maxima are parallel to the rolling direction. The optimum properties of Cunico are developed by a process consisting of cold working followed by reheating and rapid quenching from the annealing temperature, and a subsequent aging treatment. The strains induced by the precipitates in both alloys make boundary motion more difficult and result in increased H_c. These precipitates are also considered to be single domains which increase the H_c.

The Fe-Co-Mo alloys also show variations in magnetic properties as a result of thermal treatments which control the size and distribution of the precipitates. They are more machinable and less sensitive to warping or cracking than the cobalt steels shown in Table 9-4.

Table 9-6
TYPICAL PROPERTIES OF SOME COMMON MAGNET ALLOYS

Trade name	Nominal composition (wt %)	Remanence B_r (G)	Coersive force H_c (Oe)	Approximate Curie temperature (°C)
Cunico	50 Cu, 21 Ni, 29 Co	3,400	700	0.80
Cunife	60 Cu, 20 Ni, 20 Fe	5,400	500	1.20
Comol	12 Co, 17 Mo, Fe	10,000	240	1.00
Remalloy	12 Co, 20 Mo, Fe	9,000	320	1.40
Alnico				
I	12 Aℓ, 20 Ni, 5 Co, Fe	7,100	425	1.35
II	10 Aℓ, 17 Ni, 12.5 Co, 6 Cu, Fe	7,200	545	1.65
III	12 Aℓ, 25 Ni, Fe	6,800	460	1.35
IV	12 Aℓ, 28 Ni, 5 Co, Fe	5,500	700	1.30
V	8 Aℓ, 14 Ni, 24 Co, 3 Cu, Fe	12,500	600	5.00
VI	8 Aℓ, 15, Ni, 24 Co, 3 Cu, 1.25 Ti, Fe	10,300	750	3.65
Barium ferrite	$BaO \cdot 6 Fe_2O_3$ ($BaFe_{12}O_{19}$)	2,200	1,800	1.00
	Oriented $BaO \cdot 6 Fe_2O_3$	3,800	2,000	3.25
Vectolite	30 Fe_2O_3, 44 Fe_3O_4, 26 Co_2O_3	1,600	900	0.50

From Lyman, T., Ed., *Magnetic, Electrical and Other Special-Purpose Materials*, Vol. 1, 8th ed., American Society for Metals, Metals Park, Ohio, 1961, 790. With permission.

The Alnico group of alloys (Fe-Aℓ-Ni-Co) are brittle and cannot be mechanically worked. These are available in the sintered or cast form. The mechanical properties of the sintered alloys are much higher than those which are cast because of the small particle size.

The high coercive forces of the Alnico alloys result from the fact that they also depend upon the presence of two phases. The presence of finely divided, well-dispersed, second-phase precipitates distorts the lattice and restrains the motion of the domain walls. The single domain nature of the fine precipitates with high H_c are primarily responsible for their high energy products and high H_c. When precipitated particles are small enough, they consist of a single domain and, having no domain walls, demagnetization must occur by the simultaneous reversal of all the spins. This requires high fields and increases the H_c of the alloy.

A region exists in the Fe-Aℓ-Ni system in which two particular phases are present. One of these, α, is a disordered BCC solid solution. The other phase, α', is ordered and has a composition given by $Fe_2NiA\ell$. Upon proper heat treatment, the α phase precipitates in the α' phase. The degree of fineness and uniformity of this precipitation determines the number and extent to which the precipitating particles can exist as single domains as well as inhibit the motion of the domain walls. The finer and more evenly distributed that the α phase is in the α' matrix, the greater will be the contribution to the coercive force of the alloy and to the resistance to the motion of the domain walls. Optimum precipitation conditions for this occur at the beginning of precipitation where the particles have had just enough time to precipitate, but insufficient time to grow and coalesce. The presence of cobalt has the effect of increasing H_c and the energy product.

These alloys also respond to magnetic annealing. When the thermal treatment is carried out in a magnetic field the energy product and coercive force are increased. Details of the thermal and magnetic treatments are given below.

The solution treatment consists of heating to 1200 to 1300°C followed by controlled cooling to about 500°C, then cooling to room temperature. The alloys are then reheated for precipitation to about 600 to 650°C. Magnetic fields of about 1000 to 3000 Oe are applied during the cooling of Alnico V and VI from the high-temperature treatment. This treatment increases the maximum energy product for alloys contain from 7 to 11 Wt % Aℓ; those with 8 Wt % Aℓ are increased by a factor of about 2.5.

Alnico alloys I, II, III, and IV show isotropic magnetic properties. Alnico V and VI are strongly anisotropic. Alnico V, the strongest magnetic material of this class, has its highest magnetic properties parallel to the magnetic field applied during thermal treatment. The magnetic properties in the other directions are poor. This anisotropy must be considered in magnet design. It can be advantageous to apply the field in the desired directions during the manufacture of such magnets for specific applications.

Because of its high energy product, Alnico V is the most used of these alloys. Alnico III is used primarily for magnets of small section size.

Alnico II may be used in preferences to Alnico alloys I, III, and IV because of its higher energy product. Alnico IV is used where demagnetizing conditions may exist because of its high coercive force.

Mechanical shock and aging phenomena at ordinary temperature have little effect upon the properties of Alnico alloys. Extraneous magnetic fields and large variations in temperature can cause changes in the magnetic properties of these alloys. The prior exposure of these magnets to a field higher than that of any extraneous field anticipated in service will stabilize the Alnico alloy. The effects of ordinary temperature fluctuations are usually negligible.

9.14.1.1. Barium Ferrites

The barium ferrites are the most important type of this class of "ceramic" permanent magnet materials. They possess much higher coercive forces than the Alnico alloys. These ferrites are between two and three times more resistant to demagnetization than are the Alnico types. In addition, the electrical resistivities of the ferrites are greater than those of the Alnico alloys by a factor of 10^6 at least.

These ferrites are composed of BaO and Fe_2O_3. Data for such a compound are given in Table 9-5 for both the random and oriented conditions. Other compounds of this class are formed by the substitution of MO for the BaO, in which M usually is a divalent metal from the first transition period in the Periodic Table.

These barium-base compounds have hexagonal symmetry and, consequently, are strongly anisotropic. Differences in the hexagonal structures result in different sublattices and in the magnetic interactions between them. This results in materials with large ranges of magnetic anisotropy energies and saturation magnetizations. The easy direction of magnetization of some of these compounds lies in their basal planes.

The magnetic anisotropy increases the utility of these ferrites as permanent magnet materials (see Sections 9.3 and 9.7). The anisotropy is produced by orienting the powder particles prior to further processing. This can be done by placing the loosely packed powders within a mold and placing it in a strong magnetic field. The particles tend to align themselves with their easy directions of magnetization parallel to the applied field. They are then pressed and sintered into the desired configurations. The particles also may be oriented during pressing. The orientation of the particles is unaffected by the sintering.

The particle size affects the magnetic properties of the ferrites. Very small particles have high coercive forces and remanences of about half the saturation magnetization. This is further evidence for considering the small particles to be single domains and indicates that domain rotation takes place. Where the particle size is larger, the coercive force decreases and the remanence rises. This indicates that domain walls are present and move. The manipulation of the grain size by controlling the original particle

Table 9-7
TYPICAL PROPERTIES OF SOFT MAGNETIC ALLOYS

Trade name	Nominal composition (wt %)	Maximum permeability	Remanence B_r (G)	Coercive force H_c (Oe)
Thermenol	16 Aℓ, 3.5 Mo, Fe	60,000	2,070	0.018
16 Alfenol	16 Aℓ, Fe	80,000	4,000	0.044
Sinimax	43 Ni, 3 Si, Fe	50,000	5,500	0.06
Monimax	48 Ni, 3 Mo, Fe	60,000	8,900	0.06
Supermalloy	79 Ni, 5 Mo, Fe	300,000 min	4,000 min	0.006
Mumetal	77 Ni, 5 Cu, 1.5 Cr, Fe	100,000	2,300	0.30
Supermendur	49 Co, 2 V, Fe	70,000	21,400	0.23
	35 Co, 1 Cr, Fe	10,000	11,000	0.63
Ingot Iron		5,000	7,700	1.00
	0.5 Si-Fe	3,000		0.90
	1.75 Si-Fe	5,000		0.80
	3.0 Si-Fe	8,000		0.70
	Oriented 3.0 Si-Fe	50,000	12,000	0.09
	Oriented 50% Ni-Fe	150,000	14,500	0.09
	50% Ni-Fe	100,000	9,000	0.05

From Lyman, T., Ed., *Magnetic, Electrical and other Special-Purpose Materials*, Vol. 1, 8th ed., American Society for Metals, Metals Park, Ohio, 1961, 782. With permission.

Table 9-8
TYPICAL ALLOYS OF NICKEL AND COBALT FOR SOFT-MAGNET APPLICATIONS

Nominal composition (wt %)	Maximum permeability	Remanence B_r (G)	Coercive force H_c (Oe)	Approximate Curie temperature (°C)
45 Ni, Fe	30,000	8,000	0.20	440
47 to 50 Ni, Fe	50,000	8,000	0.07	535
50 Ni, Fe	70,000	8,000	0.05	535
Oriented 50 Ni, Fe	50,000—100,000	14,500	0.10 to 0.20	535
48 Ni, 3 Mo, Fe	35,000		0.10	
79 Ni, 4 Mo, Fe	100,000	5,000	0.05	421
79 Ni, 5 Mo, Fe	800,000	5,000	0.005	
77 Ni, 5 Cu, 1.5 Cr, Fe	100,000	3,000	0.05	
50 Co, Fe	5,000	14,000	2.00	
49 Co, 2 V, Fe	4,500	14,000	2.00	526
35 Co, 1-2 Others, Fe	10,000	13,000	1.00	

From Lyman, T., Ed., *Magnetic, Electrical and Other Special-Purpose Materials*, Vol. 1, 8th ed., American Society of Metals, Metals Park, Ohio, 1961. With permission.

size distribution, by heat treatment, or by other means, thus provides methods for the intentional modification of the properties of ferrites.

These ferrites are magnetized by applying large external fields. A "rule-of-thumb" is that the external field should be at least five times the expected coercive force of the ferrite. Their demagnetization is accomplished most easily by heating them above the Curie temperature (about 510°C).

One very large application of the barium ferrites is for focusing magnets for television tubes. They also have wide use in small d-c motors and in compact torque drives, where space is limited, and where high energy product magnets, resistant to demagnetization, are required.

Many ferrites other than those based upon BaO are available. Two of the more important of these are $PbO \cdot 6Fe_2O_3$ and $SrO \cdot 6Fe_2O_3$. They have high maximum energy products and high coercive forces. These, too, may be made to have isotropic or oriented magnetic properties.

Vectolite (Table 9-6) is a mixture of oxides. It is virtually an electrical insulator. This material is cooled in a magnetic field after sintering. Its maximum properties in the oriented direction are given approximately as: $B_r \simeq 1900$ G, $H_c \simeq 1100$ Oe, and $(B_dH_d)_{max} \simeq 0.8 \times 10^6$ G-Oe. These increase as the density increase; the data given in the table are typical values. This combination of properties has led to its use in frequency applications.

9.14.2 Soft Magnetic Materials

As previously noted (Section 9.14) magnetically soft materials have hysteresis loops with small areas. The values of B_s are high and those for H_c are very small (see Tables 9-7 and 9-8).

Impurity elements such as carbon, sulfur, nitrogen, and oxygen have deleterious effects in alloys because they distort the crystal lattice and hinder the easy formation of domains and their motion. Higher concentrations of these elements result in the precipitation of particles within the grains of the iron and cause similar effects. Because of this, the carbon content of these alloys is kept as low as possible, 0.005 Wt %C being preferred with a maximum of 0.01 Wt %C.

The resistivity of pure iron is too low for its application to alternating-current circuits which constitute the largest application of soft magnetic materials. The eddy current losses are too high. The elements silicon or aluminum are added to form substitutional solid solutions which largely overcome this difficulty by increasing the resistivity. In so doing, they decrease B_s. The use of these elements is limited because alloy contents in excess of about 4.5 Wt % silicon or 8 Wt % aluminum cause the alloys to become brittle. In addition, both of these alloying elements confine the alpha-to-gamma allotropic transformation to narrow loops which extend over a range of about 2 Wt % beyond the iron terminus. Thus, most commercial alloys of these kinds are BCC in structure because large grain sizes can be produced easily by avoiding the transformation. This decreases H_c because fewer grain boundaries offer less resistance to the motion of domain walls.

As noted in Section 9.3, the magnetic properties of iron and its alloys are anisotropic, with the easy direction of magnetization being [100] (Figure 9-8). Silicon-iron alloys are mechanically and thermally processed to align the grains in such a way as to have their crystal orientations such that the cube faces are oriented parallel to the direction of rolling. This provides a material with magnetic properties superior to those of a comparable material which is not oriented. The grain size of the oriented material is kept as large as possible to minimize grain boundaries.

Iron-silicon alloys have more desirable magnetic properties than pure iron, except for the decrease in B_s. These alloys vary in composition from 0.5 to 4.5 Wt % silicon. Most of the commercial alloys are oriented. The value of B_s of oriented material is about twice that of nonoriented material. Iron-silicon alloys, free of the impurity elements noted previously, do not show significant aging. Therefore, hysteresis losses do not usually increase with time at the temperature of operation.

The embrittling effect of aluminum limits its substitution for silicon. Fe-Si-Aℓ alloys have desirable magnetic properties at low flux densities. Such alloys approximate the magnetic properties of Fe-Ni alloys.

The effect of nickel on the permeability of Fe-Ni alloys reaches a maximum at about 80 Wt % Ni. The energy products of these alloys have maxima in the neighborhood of about 50 Wt % Ni. The nickel content of most of the commercial alloys ranges between 40 and 60 Wt %. Nickel contents close to 50% give the highest saturation values. Other alloying elements are added to the Fe-Ni base for specific purposes.

Iron-cobalt alloys have saturation values higher than that of pure iron. These alloys may contain up to about 65 Wt % Co. The maximum saturation value occurs in the alloy containing 34.5 Wt % Co. Alloys containing more than about 30 Wt % Co are brittle. Small additions of chromium and vanadium can minimize this tendency, when such alloys are suitably processed.

The Fe-Si alloys are used in the form of thin sheets from which laminated components are made. This is done to decrease eddy-current losses. These losses decrease with increasing silicon contents; the permeabilities remain high, however. Some typical applications include pole pieces for a-c and d-c electric motors, cores for transformers, high-efficiency distribution and power transformers, as well as large generators.

The alloys noted in Tables 9-6 and 9-7 represent only a few of the commercially available, soft magnetic alloys. Those containing higher amounts of alloying elements generally possess high permeabilities and show low losses. They find wide application in such components as audio transformers, coils, relays, magnetic amplifiers, high-frequency coils, magnetic shields, and d-c electromagnets.

Ferrous alloys containing 50 to 90 Wt % Ni are mechanically worked and then annealed in magnetic fields to provide oriented sheets. These alloys have high permeabilities and low H_c. Some of these alloys have rectangular hysteresis loops (Figure 9-15b). Other alloys whose compositions fall within the ranges 15 to 25 Wt % Fe, 15 to 70 Wt % Co, and 10 to 70 Wt % Ni have similar properties. The narrow rectangular loops result in essentially constant, high permeabilities. This means that virtually constant high values of B are obtained for small H and are maintained over wide ranges of H. These alloys are used in such devices as sensitive relays, audio transformers, and amplifier coils, to give only a few examples.

9.14.2.1. Ferrites

Ferrites for soft magnetic applications have the composition and structures described in Section 9.12. Materials of this class are very useful at high frequencies because their resistivities are at least of the order of 10^6 greater than those of the high-permeability alloys discussed previously. Many of these materials have low values of B_s and T_c. They also have low coercive forces and, consequently, very narrow hysteresis loops. This combination of properties minimizes eddy curren losses. This eliminates the necessity for laminations, and minimizes hysteresis losses. The hysteresis losses, which increase as a function of frequency, are smaller for the ferrites at high frequency than for the soft magnetic alloys. In addition, ferrites have high relative permeabilities so the volume of a suitable ferrites for a fixed magnetic effect will be smaller than that required of a soft magnetic alloy. On the other hand, the saturation magnetizations of this class of ferrites are much smaller than those of the soft magnetic alloys. This confines the application of many ferrites to high-power, high-field uses such as power transformers, generators, and motors.

Other ferrites are available which possess "rectangular" hysteresis loops (Figure 9-15b). These have high saturation values induced by relatively small fields. During magnetization which occurs in the vertical portion, a magnetization front exists such that the material is magnetized in opposite directions on either side of the front. At saturation magnetization the material is magnetized in one direction.

Still other ferrites are used as components of electronic filters, microwave devices, magnetic switches, and "memory" elements for computers. The application of ferrites in memory and switching devices is made in the form of very thin films (about 10^{-5} cm) and tapes. This limitation is imposed to prevent the existence of more than one domain across the thickness of the film. The reversal of magnetization, thus, is con-

Table 9-9
PROPERTIES OF SOME MIXED FERRITES

Ferrite	Coercive force H_c(Oersted)	Saturation magnetization B_s(Gauss)	Initial permeability
50-50 Mn-Zn	0.1	200	2000
50-50 Mg-Mn	0.5	215	—
50-50 Mg-Zn	0.1	207	500
30-70 Ni-Zn	—	320	80
40-60 Cu-Zn	0.5	—	1100
40-60 Cu-Mn	1.0	230	—

After Chikazumi, S., *Physics of Magnetism,* John Wiley & Sons, New York, 1966. With permission.

fined to domain rotation, eliminating domain wall motion, and decreases the time for imprinting bits of information.

As would be expected from their bonding and structures (Section 9.12), ferrites are hard and brittle. Most bulk ferrites are compacted and sintered in, or very close to, their final configurations because of the difficulty of machining these materials. Any machining is done by grinding.

Unlike the soft magnetic alloys, the compositions of most ferrites have not been standardized. The properties of some ''mixed'' ferrites are given in Table 9-9.

The ferrites given in the table are ''mixed'' ferrites because two ions are substituted for the M^{+2} ion in the $M^{+2}Fe^{+3}O_4$ molecule. The name Ferroxcube has been given to some of the double ferrites in the table. As previously noted, materials such as these are used because of their small energy losses in high-frequency fields and because of their low eddy current losses. Their permeabilities also are important.

It should be noted that the properties of ferrites vary greatly with composition, powder processing, compacting, and sintering regimes.

9.15. PROBLEMS

1. Use the Weiss theory to explain how a ferromagnetic material can show no magnetism at temperature below the Curie temperature.
2. Use Figure 9-1 to obtain a graph of the magnetization as a function of temperature.
3. Explain, in nonmathematical terms, why the Curie temperature should be proportional to the Weiss constant.
4. Provide a possible explanation for the nonlinear behavior of $1/\chi_P$ vs. T in the range just above T_c.
5. Make use of the magnetic component of heat capacity to explain the thermoelectric properties of nickel and nickel-base thermocouple elements at temperatures at and below T_c.
6. Explain how the linear coefficient of expansion and the thermal conductivity of ferromagnetic alloys are expected to behave at and below T_c.
7. Make an approximate calculation for J_e for nickel if T_c is 360°C.
8. Why are mechanisms involving magnons valid only at low temperatures?
9. Devise a simple apparatus for the detection of the Barkhausen effect.
10. Explain why large numbers of domains decrease the magnetic energy of a ferromagnetic solid.

11. Why should the extent of a Bloch wall be determined by the stored energy involved?
12. Explain the utility of B vs. H curves in engineering applications.
13. Deduce the number of unbalanced spins on the Ni ion from the Ni-Aℓ system.
14. Why should Pd have such a small effect upon the magnetic properties of the Ni-Pd system?
15. Discuss the reasons for the importance of crystalline perfection in Heusler-type alloys.
16. Assume the magnetization curves for the sublattices as given in Figure 9-21, and graph the magnetization curves for the ferrites in which $x = 0.5$ and 1.5.
17. Outline the essential differences between diamagnetism, paramagnetism, ferromagnetism, ferrimagnetism, and antiferromagnetism in terms of the basic phenomena involved.
18. Derive Equation 9-60 as indicated in the text.
19. Show the area enclosed by the B vs. H curve may be expressed in terms of energy.
20. Why should the maximum energy product be considered as being a useful criterion for the selection of permanent magnet materials?
21. Contrast the magnetic properties desired in both hard and soft magnetic materials. Give the reasons for their desirability in each case.

REFERENCES

1. **Dekker, A. J.,** *Solid State Physics,* Prentice-Hall, Englewood Cliffs, N.J., 1957.
2. **Chikazumi, S.,** *Physics of Paramagnetism,* John Wiley & Sons, New York, 1966.
3. **Kittel, C.,** *Introduction to Solid State Physics,* John Wiley & Sons, New York, 1966.
4. **Stoner, E. C.,** *Magnetism,* Methuen, New York, 1936.
5. **Bates, L. F.,** *Modern Magnetism,* Cambridge, New York, 1961.
6. **Martin, D. H.,** *Magnetism in Solids,* London Iliffe Books, 1967.
7. **Berkowitz, A. E. and Knelles, E.,** *Magnetism and Metallurgy,* Vol. 1, Academic Press, New York, 1969.
8. **Bozorth, R. M.,** *Ferromagnetism,* Van Nostrand, New York, 1951.
9. **Anderson, J. C.,** *Magnetism and Magnetic Materials,* Chapman and Hall, London, 1968.
10. **Morrish, A. H.,** *The Physical Principles of Magnetism,* John Wiley & Sons, New York, 1965.

APPENDIX A

USEFUL PHYSICAL CONSTANTS

Constant	Symbol	Value: CGS	Value: SI
Electron rest mass	m	9.11×10^{-28} g	9.11×10^{-31} kg
Electron charge	e	4.80×10^{-10} esu	1.60×10^{-19} coul
Planck's constant	h	6.63×10^{-27} erg sec	6.63×10^{-34} J sec
Planck's constant/2π	$\hbar$	1.05×10^{-27} erg sec	1.05×10^{-34} J sec
Boltzmann's constant	k_B	1.38×10^{-16} erg/K	1.38×10^{-23} J/K
	k_B	8.63×10^{-5} eV/K	
Electron volt	eV	1.60×10^{-12} erg	1.60×10^{-19} J
Electron volt/molecule	eV/a	23.06 kcal/mol	
Gas constant	R	1.987 cal/(mol K)	8.31 J/(mol K)
Avogadro's number	N, N_A	6.02×10^{23}/mol	6.02×10^{26}/kmol
Atomic mass unit	amu	1.66×10^{-24} g	1.66×10^{-27} kg

APPENDIX B

CONVERSION OF UNITS

Unit	Symbol	Conversion operation	Resulting units
Electrical potential (volt)	V	V/300	statvolt
Electrical current (amp)	A	$A/(3 \times 10^9)$	statamp
Electric field (V/cm)	$\bar{E}$	$\bar{E}/300$	statvolt/cm
Conductivity [$(\Omega\text{-cm})^{-1}$]	σ	$\sigma \times 9 \times 10^{11}$	esu conductivity
Resistivity (Ω-cm)	ϱ	$\varrho/(9 \times 10^{11})$	esu resistivity
Mobility [$cm^2/(V\ sec)$]	μ	$\mu \times 300$	cm^2/statvolt sec
Current density (amp/cm^2)	j	$j \times 3 \times 10^9$	statamp/cm^2
Magnetic flux density (Weber/m^2)	T	$T \times 10^4$	gauss
Magnetic field strength (amp turns/m)	H	$H \times 4\pi \times 10^{-3}$	oersted
Thermal conductivity [cal/(cm sec°C]	$\varkappa$	$\varkappa \times 422$	Watt/(m°K)

INDEX

A

B

C

D

E

F

G

H

I

J

K

L

M

  density, 31, 138, 155, 160—161, 165

N

O

P

Palladium, properties of, 40, 57—58, 76, 114, 142
Palladium—gold systems, 53, 58, 64

Q

R

S

T

U

V

W

X

Y

Z